DUE DATE

AUG 2 8 1991			
APR - 9 1992			
APR - 9 1992			
APR 2 3 1992			
MAR 2 4 1997			
	201-6503		Printed in USA

Power and Process Control Systems

For the Plant Engineer and Designer

Michael J. G. Polonyi

McGraw-Hill

New York St. Louis San Francisco Auckland Bogotá
Caracas Hamburg Lisbon London Madrid
Mexico Milan Montreal New Delhi Paris
San Juan São Paulo Singapore
Sydney Tokyo Toronto

Library of Congress Cataloging-in-Publication Data

Polonyi, Michael J. G.
  Power and process control systems / Michael J. G. Polonyi.
    p.   cm.
  Includes index.
  1. Electric power systems—Control. 2. Electric power—plants—
Control. I. Title.
TK1005.P666 1991   621.31′7—dc20   90-43289

1 2 3 4 5 6 7 8 9 DOCDOC 9 6 5 4 3 2 1

ISBN 0-07-050414-8

The editors for this book were Robert W. Hauserman, Caroline Levine
and George F. Watson. The designer was Naomi Auerbach and the
production supervisor was Thomas G. Kowalczyk. It was set in
Century Schoolbook by McGraw-Hill's Professional Publishing
composition unit.

To My Friends and Family

Contents

Preface

The purpose of this book is to establish a common ground between control theory as a tool and the realities of power systems and plant operation and design. It is intended for all audiences. Those who do not have formal—advanced—power system engineering training may skip the math portions without significant loss of continuity. It may prove particularly useful for both those involved in the daily operation of power plants and those involved in their design. There is a need that idiosyncratic features be accounted for at the design level, since it is much harder to implement them once the power plant is in operation.

It is my experience that, in general, not enough attention has been devoted to the dynamic response requirements and possibilities of the major components that affect a system's behavior—at least not in a way that is both practical and acceptable to plant and system engineers and designers alike.

Chapters 1, 2 and 7 are strictly tutorial, both from a theoretical and from a practical point of view. Chapter 1 uses level and temperature control examples to establish some working concepts that are maintained throughout the book. Chapter 2 describes instrumentation in general and the role it plays in closed-loop control.

An original set of equations for PID-tuning based on standard forms developed by Graham and Lathrop, which can be used for any application using PID-type controllers, is presented in Chap. 3. Special emphasis is placed on control power supply requirements, a subject hardly mentioned in control literature. Chapter 4 also explains why the "cost" of control is not free.

A series of "tools"—models and equations—is presented to help readers understand and solve basic problems, such as PID controller tuning for boiler drum level (Chap. 5) and steam pressure control (Chap. 4).

Chapter 6 describes cogeneration from a dynamic control point of view and how it must balance—and compromise—with both the requirements from the utility's point of view and the local energy (steam and electric) demand it is trying to meet.

Chapter 7 describes the major operational factors in large power systems and how they interact with each other. Chapter 8 is dedicated to power-frequency control and the factors affecting its stability. An original table illustrates and compares the flexibility and response of conventional power plants.

Chapter 9 illustrates the transient energy response of fossil and nuclear power plants and highlights the major dynamic influence that steam as a power source really is. Chapter 10 is dedicated to stability concepts of interconnections. An original feature of this chapter is the inclusion of a proposed coordinated alternator excitation control scheme to avoid drift and minimize circulation currents to improve stability.

The book closes with an appendix, which includes a collection of instrumentation, control, and logic diagrams for a solid-waste burning plant. These diagrams can be adapted as useful models for other boiler plants.

Michael J. G. Polonyi

Power and Process
Control Systems

Introduction to Process Control

State of the Art

One of the first assignments I chose as a power plant instrumentation and controls engineer was to find the best possible proportional, integral, and derivative (PID) controller settings for a high-pressure steam boiler drum level-control loop. This may sound like a relatively simple task.

After looking through books, articles, reference journals, catalogs, specifications, brochures, and instructions, and not finding a simple instruction, model, or equation to use straightforwardly, I finally came to the conclusion that either nobody understood level control or nobody *cared* what the settings should be, as long as the boiler held the drum level just about where it is supposed to be. Although, in many cases, the boiler does just that, it can be demonstrated that plants will operate a lot better if proper dynamic modeling, simulation, and analysis techniques are used routinely during and after plant engineering and design.

Process control literature in general uses level control as a typical introductory example and tries to make the case that process control modeling is useless as a practical tool 99 percent of the time.

This is not so. The importance of PID control is by no means a closed issue, even though most texts treat the subject marginally. PID control is a time-domain problem, and this does not sit well with many control systems authors, who prefer to analyze it in the frequency domain.

On the other hand, there is a historical reason why process control engineering still does not require extensive dynamic mathematical modeling in most process applications. Process control evolved from crude empirical techniques for controlling boiler level and steam engine speed governors with such devices as mechanical levers, floats, and flyballs.

Although equipment specifications have become much more accurate, the built-in time lags of the equipment are still large enough to allow empirical controller tuning without regard to time lags, with no major complications. "When everything else fails, go over to manual control" is the motto of many plant operators, who will readily open an automatic control loop and exercise manual control if they feel it is safer than leaving it to the automatic controller. This reflects the low level of reliability that closed-loop control provides.

This brings up another interesting aspect of control engineering, the fact that a multimillion dollar piece of equipment, like a high-pressure steam boiler, an airplane, or a space shuttle, depends on a relatively inexpensive "box" for control. The irony is that the box—i.e., computer—is not itself as reliable—functionally—as the equipment it is controlling!

In the end, process control proves to be an elusive and intuitive science/art. It requires a background in almost all engineering disciplines: mechanical, electrical, process chemical, and computer science. Moreover, the laws of physics and chemistry lose their power when one tries to understand transient phenomena in a practical way. Somehow, one has to forget previously acquired knowledge and try to find root causes on the basis of two mathematical abstractions: *gain* and *time lags*. Once a closed-loop control process is under way, the parameters of the different components—a blend of electrical, mechanical, chemical, and computational—of the system are shared, and the observed behavior may have a root cause, or combination of causes not immediately apparent. Some dynamic parameters, such as gain and dead time, are *shared* by the whole loop, regardless where in the loop they originate. But *time-constant placement* has a crucial effect on the stability and the magnitude of the response.

Control systems theory, even though it is so highly developed and sophisticated, fails to deal with the need to establish a level of feasibility and reliability that a model entails. In other words: it becomes "educated guessing." This is not necessarily a bad thing, but the problem is that starting from a poor understanding of a process may lead to uncontrollable designs, and the uncontrollability may not be immediately obvious to the designer. In other words, the design process itself becomes unstable!

Control theory is a tool, and engineers should learn to use it properly *through practical applications*. Otherwise, using it is like taking the subway without knowing where to get off, which inevitably leads nowhere, usually at great expense and loss of time.

The suggestions and recommendations presented here are based on a combination of process control theory and the author's several years' field experience in fossil power plant operation, load dispatch, instrumentation maintenance, and power plant engineering and design.

Most of the tests described in later chapters were run on water-tube boilers of the *9 de Julio* power plant in Mar del Plata, Argentina. Most of the design engineering was done with Gibbs & Hill, Inc. New York. Other important sources of experience were the AYEE National Load Dispatch System in Argentina, and, in the United States, the Columbia University Department of Nuclear Engineering and several cogeneration projects.

Level Modeling Example

Consider level control as an example. How would we dynamically describe, in terms of a mathematical model, the level of a tank partly full, with fluid both coming in and going out (Figs. 1.1 and 1.2)?

The problem sounds simple enough. The change in level is proportional to the fluid coming in minus the fluid going out:

$$\frac{l_{t1} - l_{t2}}{t_1 - t_2} \propto \text{fluid in} - \text{fluid out}$$

where l_{t1} is the level at time t_1 and l_{t2} is the level at time t_2.

How do these variables relate *dimensionally*? Water in $q_{(t)}$ minus water out $q_{o(t)}$ is equal to the surface area of the tank times the first derivative of the level with respect to time (dl/dt), times the density of the fluid:

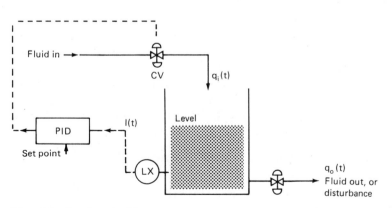

Figure 1.1   Level control loop.

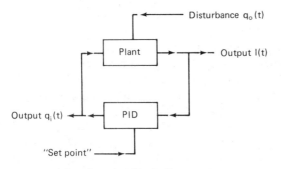

Figure 1.2   Level control block diagram.

$$Q_{i(t)} - Q_{o(t)} = \text{density} \times \text{surface} \times \frac{dl}{dt}$$

This kind of process is known as an *integrator*.

The fluid coming in can be controlled directly by a control valve (CV) or variable-speed pump at the inlet. The fluid going out—it could be steam at this point—cannot be controlled directly, however, because downstream factors control the outlet valve.

Question   How should the CV valve or (pump) be controlled so that the level stays as steady as possible, regardless how recklessly the water is drawn from the tank?

Answer   Use a PID controller to measure the level error, and let its output signal control the valve—or pump—accordingly. Why a PID controller? Because it does the job.

Question   How big should the CV valve be?

Answer   Obviously, bigger than the valve downstream, otherwise the tank may end up empty!

With the PID controller, there is now a closed, or feedback, loop. The next step is to set three PID controller parameters: proportional band, integral action, and derivative action. These settings tell the PID controller how much to respond to an input.

The proportional band concept was used in the past when control was done by levers, but it is cumbersome to use and difficult to understand. Nowadays engineers think of proportional band in terms of its inverse, gain G, which is a multiplying constant:

$$\text{Output} = \text{input} \times \text{gain}$$

A gain adjustment is provided in a control loop to reduce the excursions of the controlled variable each time there is a disturbance.

Integral action (Fig. 1.3) increases or decreases continuously the output of the controller as long as there is an error input present (**Fig.**

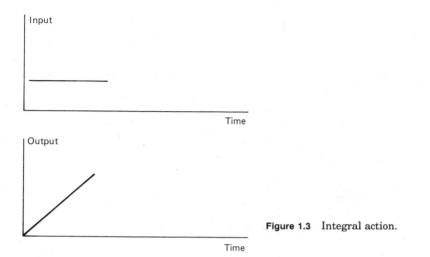

Figure 1.3   Integral action.

1.3). The rate of change is the magnitude of the input over the integral action time constant T_i. Integral action is used to eliminate steady-state differences in the set point, after a disturbance. The inverse of T_i is called *reset*. For integral action,

$$\text{Output} = \frac{\text{input} \times \text{time}}{T_i} = \text{input} \times \text{reset} \times \text{time} = \frac{1}{T_i} \int_0^\infty i(t)dt$$

Integral action makes it possible to eliminate the difference between process input and output, once the system reaches a steady-state condition.

Derivative action keeps the controller output constant for as long as a constant change in the input is present (Fig. 1.4). The magnitude of the output is the rate of change of the input times the derivative action time constant T_d. It is used to anticipate the effect of unwanted disturbances, thereby reducing their effect. T_d is also called *rate*. For derivative action,

$$\text{Output} = \frac{\text{input}}{\text{time}} = \frac{T_d\,di(t)}{dt}$$

The anticipatory effect of derivative action reduces the time it takes to reach a new condition after a disturbance, if there is enough control power to do so, which is not always the case.

Of course, input or output can change only until it reaches the physical limit of the devices. This limit is called *saturation* and the saturation effect is called *clipping* (Fig. 1.5).

We can now develop a mathematical model based on parameters

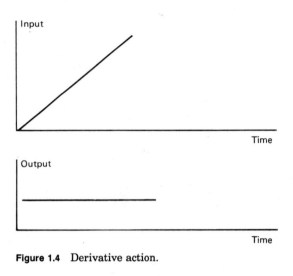

Figure 1.4 Derivative action.

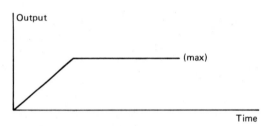

Figure 1.5 Saturation or "clipping."

that can be recognized and quantified—derivatives, gain, etc. A first cut at a block diagram is in Fig. 1.6. The set point establishes the desired level; since it is not changed often, level set-point *dynamic behavior* can be ignored. There are two inputs, and the one that matters most is the one we do not know much about: flow out, determined by events downstream. Although the tank is delivering downstream, equipment downstream provides an input. We call this an *unwanted input* or *disturbance*. A disturbance in the level ontrol loop can be re-

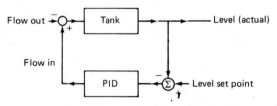

Figure 1.6 Level/flow control loop block diagram.

garded as merely a change of set point of a flow control loop, since we have to maintain the same amount of fluid coming in as going out. We call the level controller a *regulator* and the flow controller a *servo-mechanism*, or simply a *servo*. *We can find both a regulator and a servo input/output (I/O) pair in any feedback loop.*

In control theory terms,

$$\frac{\text{Level}}{\text{Water out}} = \lim_{t \to \infty} \frac{\text{output}}{\text{disturbance}} = 0 \qquad \text{regulator}$$

$$\frac{\text{Water in}}{\text{Water out}} = \lim_{t \to \infty} \frac{\text{output}}{\text{input}} = 1 \qquad \text{servo}$$

Now we are ready to give a mathematical expression to each block in Fig. 1.6 so that we can predict what is going to happen. We invoke functional analysis and control systems theory again and start filling in the blanks! This may seem a daunting prospect. One may wonder how to fill in the blanks for a control valve, a flow meter, a level meter, and a tank. One may also wonder how to eliminate dimensions. A dimensionless model is needed because the controller and control theory do not take dimensions into account; they are based only on gain and time!

If we know what the control valve is doing, we do not need a flow meter. But the level must be measured if the controller is to do the job; therefore, we need a level meter [a *level transmitter* (LX) in process control language]. We will show that dimensionless expressions are not difficult to develop.

Now the problem has been reduced to finding gain and time constants for the tank, valve, and level transmitter. Consider the tank first. What is the *integration time constant* (or simply *integration constant*) T for a tank level control application? The answer may not be obvious, but there *must* be a T associated with level control, since it takes *time* for the level to change. The time constant for a tank level control application is the *normally allowed* change in level over the maximum possible flow Q_{max}:

$$T_{\text{tank}} = \frac{A(L_{\text{max}} - L_{\text{min}})}{Q_{\text{max}}}$$

where A = tank cross-sectional area
L_{max} = maximum normally allowed level
L_{min} = minimum normally allowed level

Then the instantaneous flow rate is

$$q(t) = T_{\text{tank}}\frac{dl}{dt}$$

and its fuctional transform is

$$\mathscr{L}\left[q(t)\right] = Q(s) = sT\left[\mathscr{L}(s)\right]$$

where s is the first derivative or Laplace operator.

This answer suggests a new question: Which flow is referred to by Q_{max}? The disturbance flow or the inlet control valve flow? In other words, there are *two* time constants associated with the tank (Fig. 1.7). T_c, which is associated with the feedwater control flow that fills the tank, must be smaller than T_d, which is associated with the disturbance flow that empties the tank, otherwise the design is flawed. But we already have specified, or defined, a maximum demand Q_{max} (which could be Q_o max or Q_i max). This means that the *gains* and *time constants* that we have found do not depend on the characteristics of one piece of equipment, but on how each piece of equipment relates to the rest.

Consider now the maximum normally allowed level change. Since fluid is both coming in and going out, the level $l(t)$ can fluctuate above and below the chosen level set point (SP).

Question   The maximum normally allowed level change is the difference between which high and low levels?

Answer   This is a good question and not a simple one to answer. We may want the control valve (CV) to be fully open when the level reaches a certain low and fully closed when it reaches a certain high. Therefore the difference in levels between a fully closed and fully open valve could be our normally allowed level change. However, it is the process design team that ultimately defines a normally allowed level change that includes high, low, and set point values. As for when the valve should be fully open or closed, this will depend on dynamic considerations, not steady-state ones.

Let's look at different scenarios: (1) No load (i.e., no fluid going out), (2) maximum load, and (3) 50 percent load. In case 1 the level can only rise, so we are interested in preserving the lower range—below the set point—only. In case 2

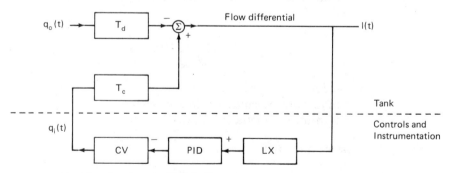

Figure 1.7   Two-flow level control block diagram.

we must try to raise the level, so we are interested in the range below the set point only, because the control valve can barely deal with the demand and restore the level at the same time. In case 3 there is enough muscle to keep the level constant without major problems. Therefore, the *range below the set point* is the only one we are interested in, to determine our control loop settings.

Question What do we mean by *allowed level change*? Does this mean we have to *choose* a maximum level change?

Answer Yes. Although this is already done by the process designer, the information is not always readily available, and we may have to redefine the ranges for our installation. This can be a tricky matter; as will be explained in the discussion of steam drum level control in a later chapter, the allowable level change should be between our high and low alarm levels.

Question Is the cross section of the tank constant?

Answer It had better be, at least over the normal level change, otherwise T would not be constant, and then we would have a characterization problem because we do not have a linear model.

Control valve G and T

The gain of the CV is 1 if we factor the maximum flow of the control valve into the tank time constant. In any case, the gain of the control valve is defined as

$$G_v = \frac{\text{maximum control valve flow}}{\text{maximum system flow}}$$

G is assumed constant throughout the operating range. However, since maximum system flow is a base value to be defined more or less arbitrarily, in practice the control valve gain can be more or less than 1, depending on the particular system.

The time constant of the control valve response will depend on the driving power of the actuator in the control valve. Typical values range from 1 to 5 s. Control valve response time constants should be kept small in comparison to the tank's T so that they can be ignored. (*Note: Small* in process control means about one-tenth the value, for all practical purposes. Present process technology does not require less, and this is one of the reasons why controller settings do not have to be highly accurate.)

Level transmitter G and T

The gain of the level transmitter (LX) will be the value corresponding to the maximum allowed level change in normal operating conditions (i.e., before the level alarm goes off) over the level transmitter range. Therefore, G is always less than 1, because the same transmitter is used for control and display purposes and because we do not want the transmitter to saturate; at least not within the allowed level range. The response time of transmitters in general can vary substantially, mainly because of the way they are installed. The response times should be kept low, if possible.

Some level-sensing systems may require a considerable time to reestablish themselves, especially after a large swing—i.e., disturbance—because the system conditions for which they were tuned, basically pressure and density, have now changed abruptly. This is the case for many level-sensing transmitters designed to be used in a high-pressure, high-temperature environment, such as steam drums and pressurized nuclear power plant water reactors. Service conditions compensation can be applied successfully.

The level transmitter gain is

$$G_x = \frac{\text{maximum normally allowed level change}}{\text{level transmitter span}}$$

Final model

Therefore, the final mathematical model looks like Fig. 1.8. The control equation is

$$\frac{L(s)}{Q(s)} = \frac{\dfrac{1}{sT}}{1 + \dfrac{1}{sT}\dfrac{G_v}{1 + sT_v} G_c\left(1 + \dfrac{1}{sT_i}\right)G_x}$$

$$= \frac{(1 + sT_v)sT_i}{s^2 T T_i (1 + sT_v) + (1 + sT_i)G_c G_v G_x}$$

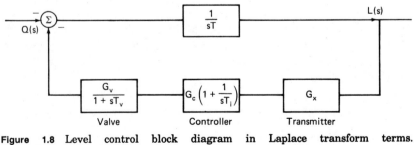

Figure 1.8 Level control block diagram in Laplace transform terms. PID $\equiv G_c(1 + 1/sT_i + sTd)$

Question What happened to the level set point input?

Answer It is no longer important. Obviously, there must be a reference set point for the level tucked into the model somewhere, but dynamically speaking it is totally irrelevant, since once we have made up our mind where the level should be, we want it to remain as constant as possible. The level set point never gets changed in this kind of application.

Optimum Response

Now that we have a model, how can we apply it? We want to find the best possible response to keep our level as close to the set point as possible. Since we have assumed that the system is already installed, we do this by adjusting the parameters in our controller: gain, reset, and rate. There will always be some deviation from the set point after a disturbance, and we want to find a setting that will minimize this.

If we want to restore the level fast, we will find more overshoot and an oscillatory response. If we want to restore the level slowly, it will take a longer time for the level to reach its set point. Obviously, we have to compromise.

Fortunately, someone has already taken the trouble to establish a choice of "best" responses. We can choose from several different modes, listed in Chap. 3. The standardized characteristic equation is

$$s^3 + a\omega s^2 + b\omega^2 s + \omega^3 = s^3 + \frac{s^2}{T_v} + \frac{sG}{TT_v} + \frac{G}{TT_vT_i}$$

where ω = natural frequency. Therefore, the best controller settings are

Loop gain:

$$\text{Gain} = \frac{bT}{a^2T_v} = G_c\,G_v G_x$$

where a and b are standard form coefficients (see Chap. 3).

Integral action:

$$T_i = abT_v$$

Derivative action:

$$T_d = 0$$

Natural frequency:

$$\omega = \frac{1}{aT_v}$$

Controller gain:

$$G_c = \frac{G}{G_v G_x}$$

As you can see, to find these settings, we need to know only equipment and process functional data, which is readily available, although this is not always the case.

To see if the level error will be eliminated, we apply the final value theorem:

$$\lim_{t \to \infty} l(t) = \lim_{s \to 0} sL(s)$$

or

$$= \lim_{s \to 0} s \frac{1/sT}{1 + \dfrac{1}{sT} \dfrac{G_v}{1 + sT_v} G\left(1 + \dfrac{1}{sT_i}\right) G_x} Q(s)$$

For a step input, $Q(s) = Q_0/s$, where Q_0 is a step magnitude. and

$$\lim_{t \to \infty} l(t) = 0$$

For a ramp input, $Q(s) = Q_1/s^2$, the error is

$$\lim l(t) = \frac{Q_1}{T} T_v$$

which means there will be a level set-point deviation as long as the load is changing.

In following chapters we will also deal with

- Steam drum level "shrink-swell" control
- Steam boiler load control modeling and optimization
- Exact PID controller settings using standard form optimization
- Dead time lag

Accuracy and Transient Travel Paths

Accuracy is a crucial parameter of all aspects of design. We must differentiate between instrument scale accuracy and process accuracy. In process control, accuracy becomes

- Gain, when making dynamic analyses
- Sensitivity or resolution, when writing sensor specifications
- Error, when calculating mass-energy balances
- Error band, when reading instruments

The degree of accuracy that is required in a process depends on the application. Mechanical and chemical processes may require very high accuracies within 0.01 percent or parts per billion. Although 5 to 10 percent is typical for the instruments' own accuracy. Electrical processes such as voltage control normally require 1 percent accuracy.

Process control must try to maintain selected variables within this accuracy while the whole system is moving to a new level of equilibrium. In this dynamic transition process, different variables may take different *travel paths*, and suddenly we are no longer dealing with just physics and chemistry, but with an abstract time phenomenon.

Although a travel path can be very strictly specified by some standard criterion while the process is in transition, it is not practical to do so because the main concern is to maintain the *transient travel path boundary constraints*. While moving between points of equilibrium, process control is therefore only loosely and qualitatively defined.

Very high sensitivity (at least of the same magnitude as the process itself), but not necessarily accuracy, is required from the instrumentation—the components that close the loop, such as sensors, transmitters, controllers, and control elements. This is necessary because the control instrumentation not only must cover a large operational range but accurately discriminate within the small band of fluctuation allowed.

When the design is inherently stable, which it should always be, PID *settings* do not need to be more than 10 percent accurate. That is because the controller *settings affect the transient travel path* strongly, but the final steady-state result only slightly.

When the controlled variable does not require more than, say, 10 percent accuracy and the rest of the system is inherently stable, control gain can be set to infinity, resulting in an on-off type controller. This suggests that a switch, relay, or other on-off devices are amplifiers of infinite gain.

The Temperature Control Problem

As another example, consider a temperature control loop (Figs. 1.9 and 1.10).

Problem The temperature $\tau(t)$ fluctuates heavily, and so does the signal to the control element.

Statement If the fluctuation of the controlled variable—temperature—were within the allowed error band, there would not be a problem.

The goal, therefore, is to reduce the amount of fluctuation in both the temperature and the signal to the mechanical control element (a

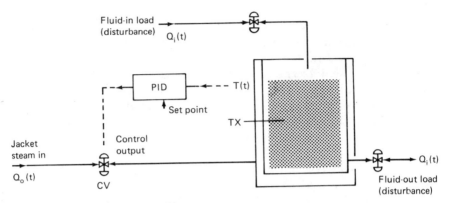

Figure 1.9 Temperature control loop.

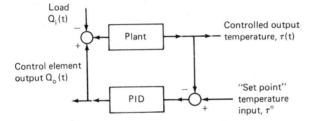

Figure 1.10 Temperature control block diagram.

steam valve), which is not designed to respond to large continuous and rapid fluctuations.

Question Why does the temperature $T(t)$ fluctuate?

Answer Either (1) the system is continuously undergoing major load upsets $Q_i(t)$ or (2) the system is unstable, i.e., it oscillates.

Just by observing a system's behavior, it is not immediately obvious whether (1) or (2) is the cause of temperature fluctuations $T(t)$, but by breaking the loop—manual control—load fluctuations can be observed. (This is not possible in a closed loop because the contribution of the control element creates loadlike responses designed to offset the true load fluctuations.)

Breaking the loop is the same as reducing the controller gain to zero. By reducing the controller gain, it is possible to discriminate between instability and load fluctuations. If the temperature fluctuations decrease, the system was unstable at that gain. If the temperature fluctuations increase, then controller action was indeed offsetting them to a certain extent.

Gain can make a system stable as well as unstable. Infinite gain makes a system unstable because the system must now oscillate between two ceilings, i.e., saturate, since nothing can grow without bound forever. A stable system, on the other hand, implies a lower deviation from the set point.

In many systems, instability cannot be tolerated even once because they self-destruct. An example is the runaway temperature condition that can occur in many types of nuclear reactors, in which temperature keeps increasing even after the system is shut down. Instances of this are bound to happen because technology is not as reliable as a natural process.

Many times controller and control element are integrated, and it is not possible to turn the loop to *manual*. This would be the case for self-regulating, safety, or relief valves.

Regardless of the cause of the temperature fluctuations, the control system is failing to do its job. Load fluctuations should not be a cause of excessive temperature swings if the system was designed to handle them. Nor should the system become unstable trying to control the output variable.

Solution for heavily fluctuating load

If the load applied to the system is within specifications, obviously a stronger control action is needed. This means a higher controller gain if the control elements have a sufficiently large power source behind them.

If the control element is close to saturation, this is an indication that there is not enough control power supply available. Operating close to saturation also causes excessive control element wear. Increasing the steam pressure and using a larger control element will reduce the burden on the control element.

Increasing controller gain to offset the output variable's fluctuations may ultimately make the system unstable; there is a limit to how much more gain the system will take. The objective is to reduce fluctuations to within the allowed error band, and this may still leave a steady-state error that cannot be eliminated by increasing the gain further. Gain by itself will not eliminate the steady-state error. To eliminate steady-state error, integral action must be added into the controller equation.

Although integral action eliminates the steady-state error, it also creates two new problems: (1) it reduces the margin of stability of the closed-loop system and (2) it introduces a second variable, which makes the search for a unique solution for the settings, or tuning the controller, much more difficult. This is discussed in Chap. 3.

Solution for an unstable system

If the system is unstable, there is too much loop gain or too much integral action from the controller. Gain can be generated by any part or component of the system, but only the controller gain can be adjusted to compensate for the overall loop gain once the system is designed.

If the control element is a thermostatic or self-regulating valve and gain adjustments are not available, and it is oversized, the system can be slowed down by restricting the rate of heating or cooling. For example, if the heating medium is steam, it can be throttled down with a manual valve upstream of the control valve.

If oscillations occur with relief or safety valves, it is necessary to increase the time response of the system to absorb some of the energy by increasing the volume of the medium. This storage process releases energy when the load increases and vice versa.

Plant Gain, Control Gain, and Characterization

In the temperature control example, the implicit assumption was made that all the dynamic parameters of the system remained constant and that the only variables were the input and the output. This assumption can be made only when load fluctuations remain small compared with the overall capacity of the system.

When the load fluctuations or disturbances remain relatively small, the system can be assumed to be linear; i.e., all the associated gains and time constants *are* constant. This is the only way to successfully apply control systems theory for PID-controller tuning.

When the disturbances are comparatively large, the system becomes *nonlinear*; i.e., gains do not stay constant, but change with the load. Eventually some parts will saturate; i.e., their gains will become zero, as in an open-loop situation.

Another factor also comes into play in temperature control: a flow of *energy* is used for control, unlike level control, where a flow of mass is used. The relation between energy and temperature is *not linear*.

To make matters even more complicated, the load that the system is trying to meet varies in two different modes: (1) small/fast and (2) large/slow changes. Usually control systems work well for certain type 1 loads but not necessarily for others. This is because the gain of

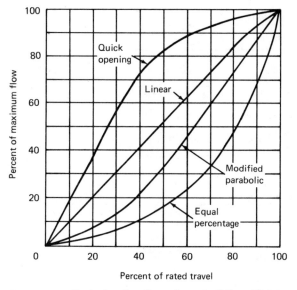

Figure 1.11   Control valve flow characteristics. (SOURCE: Fisher Controls)

the system changes with the type of load and the control elements must be *characterized* to compensate for it.

Characterization is used to maintain a relatively constant loop gain throughout the useful operating range. Sufficient additional power supply must be available at the control source to avoid saturation.

With new distributed control system (DCS) technology, a control valve signal can be accurately characterized while still in the system. However, this is not yet a common practice. To accurately determine the required characterization of a control element, the gain must be plotted for several operating conditions and valve positions. Control valves are almost always characterized on an *equal percentage* curve (Fig. 1.11), since this provides a continuous gain change response to compensate for the loss of sensitivity that occurs— as in temperature control—when the load increases. As Fig. 1.11 shows, however, there are other types of characterization which are used.

All sensors and control elements can be given some form of characterization, which amounts to providing adaptive gain. Figure 1.12 gives the best valve characteristics for control of liquid level, pressure, and flow for a variety of process conditions.

Liquid Level Systems

Control valve pressure drop	Best inherent characteristic
Constant ΔP	Linear
Decreasing ΔP with increasing load, ΔP at maximum load $>$ 20% of minimum load ΔP	Linear
Decreasing ΔP with increasing load, ΔP at maximum load $<$ 20% of minimum load ΔP	Equal percentage
Increasing ΔP with increasing load, ΔP at maximum load $<$ 200% of minimum load ΔP	Linear
Increasing ΔP with increasing load, ΔP at maximum load $>$ 200% of minimum load ΔP	Quick opening

Pressure Control Systems

Application	Best inherent characteristic
Liquid process	Equal percentage
Gas process, small volume, less than 10 feet of pipe between control valve and load valve	Equal percentage
Gas process, large volume (process has a receiver, distribution system, or transmission line exceeding 100 feet of nominal pipe volume) Decreasing ΔP with increasing load, ΔP at maximum load $>$ 20% of minimum load ΔP	Linear
Gas process, large volume, decreasing ΔP with increasing load, ΔP at maximum load $<$ 20% of minimum load ΔP	Equal percentage

Flow Control Processes

Flow measurement signal to controller	Location of control valve in relation to measuring element	Best inherent characteristic	
		Wide range of flow set point	Small range of flow but large ΔP change at valve with increasing load
Proportional to flow	In series	Linear	Equal percentage
	In bypass*	Linear	Equal percentage
Proportional to flow squared	In series	Linear	Equal percentage
	In bypass*	Equal percentage	Equal percentage

*When control valve closes, flow rate increases in measuring element.

Figure 1.12   Control valve flow characteristics. (SOURCE: Fisher Controls)

2

Instrumentation, Identification, and Control

Process Gain and Time-Constant Identification

Process control gain originates mostly in transmitters and control valves. The PID tuning method recommended in Chap. 3 requires at least approximate knowledge of system gains and time constants. This is not a problem, once the basic method for determining them is mastered.

System Time Constants

Time constant determination has remained a rather esoteric concept in process control, and its usefulness has not always been recognized. There are three basic time constants:

- First order
- Integration
- Dead time

First-order and integration time constants

A first-order time constant T is 0.632 the final value of the output when a first-order system lag is responding to a step input (Fig. 2.1). A first-order system is the same as an integrator on a feedback loop (Fig. 2.2). Therefore, although they are functionally different, the integration time constant and the first-order time constant have the same value.

Actually, an integrator is an idealization of a first-order system, be-

First-order system response   To unit step input, time const. = 1

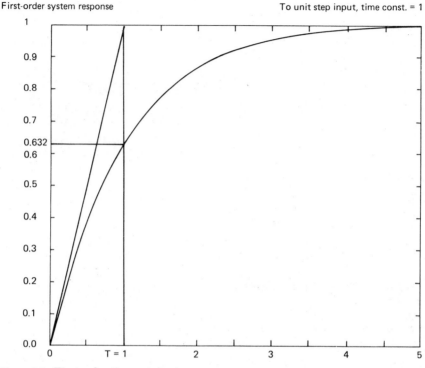

Figure 2.1 First-order time constants.

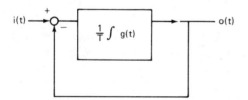

Figure 2.2 Block representation of a first-order
system: an integrator with a unit feedback loop.

cause real integrators can exist only within a certain range, otherwise
the output would grow without bound. Even if it were physically possible for the output to grow without saturating, the system would
eventually force the integrator to wind down into a first-order type of
response. Take, for example, the case of the integration time constant
of a level control application: if the level were to increase unbounded,
eventually it would stop because the system pump pressure would
equal the head of the liquid column. In other words, at some point the
system will start experiencing some feedback of its own.

Dead-time time constant

Dead time is usually introduced into the control loop by our inability to measure the development of a process variable as it occurs. For example, to determine the effect of control actions on the temperature of a fluid coming from a heat exchanger, we may have to measure the temperature at a point downstream of the process, which the product will take considerable time to reach. Another example is mixing chlorine in an open channel to kill bacteria. The residual chlorine sensor can be installed only at a point in the channel where the mixing process has already been completed, which unavoidably will create a considerable dead-time lag.

Obviously, dead time will also change if the velocity of the fluid changes, i.e., at different loads. This introduces an additional complication in our model, and the controller will have to be adaptive if we need to account for this effect.

Dead time is fairly simple to measure in a control loop, because even when it can happen in any part of the loop, it will simply add up, and we do not have to be concerned with how much of it originates where exactly. In other words, there is only one dead-time time constant per loop.

If dead time is significant, i.e., one of the three largest time constants, it must be accounted for by using appropriate equations for tuning controllers. Dead time is not hard to handle (especially with present controller technology), but it certainly adversely affects the quality of the response of the control loop.

Dead time can be observed and measured by introducing a change in an otherwise stable operation and measuring how long it takes to observe a response (Fig. 2.3).

How to recognize a time constant

Anywhere there is a time delay, there is a time constant. For example, heat exchangers have—regardless of size—typical first-order time constants between 10 to 20 s, including those in boiler furnaces and steam generators. Level control applications also typically have 10- to 20-s time constants, but they can be easily changed, since this depends on an arbitrary choice: the allowed level change.

Level *swelling* and *shrinking*, typical unstable behavior of boiler steam drums, is a difficult-to-measure parameter, with an estimated first-order time constant of 5 to 10 s. If the boiler is designed correctly, this behavior is offset by the natural tendency of the level to restore itself. (See Chap. 5 for a detailed description.)

Pressure time constants are usually much larger. For steam boilers,

Second order with dead time

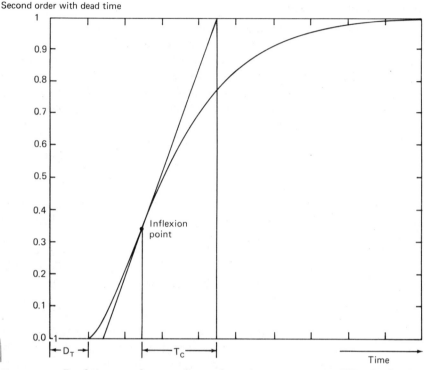

Figure 2.3 Dead-time and two first-order time constants. DT = dead time; $T_c = T_1 + T_2$; T_1, T_2 = curve time constants.

evaporator main time constants range from 50 s for once-through, to 200 to 400 s for water-tube, to 30 to 45 min for fire-tube models.

Moments of inertia, relative to the maximum load for large rotating machinery such as turbogenerators, pumps, and electric motors are also integration constants. They range typically between 2 and 20 s. The heavier the shaft with respect to the maximum load, the higher the inertia constant.

$$\text{Inertia constant} = \frac{\text{moment of inertia}}{\text{nominal power}}$$

Besides moment of inertia, electric motors and generators have several electrical time constants. Except for moment of inertia, they can be ignored when dealing with electromechanical variations. The electrical time constants usually become significant for control of voltage and reactive power flow.

Engineers who want to determine proper controller settings and system response precisely can measure main time constants with relatively simple tests if they are large enough, i.e., more than 5 s. For smaller values, measurements with electronic instruments are necessary.

Process/equipment specifications might be expected to be highly informative about time-constant values, but unfortunately more often they are not. In any case, dynamic models and time-constant information must be analyzed, the methods by which they were obtained must be specified, and the operating circumstances to which they apply must be described before equipment can be approved. Otherwise, the entire process control concept may be flawed.

Any real system—mechanical, electrical, or chemical—has many time constants. Fortunately, we are interested only in the largest ones. The higher the largest time constant, the slower the system will have to respond, which is good for the control engineer because there is plenty of time to take control action. From a process point of view, however, this may be a limiting factor since the system may not respond to fast load changes. The faster the natural response of the system, the more accurate the controller action, and settings, must be.

The natural response of a boiler, reactor, etc. depends on its physical design characteristics and on the controller settings. For example, a water-tube boiler will have a smaller main time constant than a fire-tube boiler because the water content of the first is smaller than for the second for the same steam production capacity.

Another example: To control an exothermic reaction in a batch process, temperature increases should be slow enough to allow stable heat dissipation. The allowable rate of temperature rise will depend on the chemical reaction itself and the volume of the reactor. A large volume may cause hot spots and product burning. To avoid this, the reaction itself can be slowed down by adding the reagents slowly in a large-volume reactor. The exothermic reactor time constants in this case will be

1. A time constant representing the time it takes for the temperature to reach its lower limit when addition of reactant has stopped and cooling is fully on.
2. A rising-temperature time constant when cooling is off and the reactant valve is fully open. It must be larger (i.e., slower) than the time constant in 1 for safe control.

Sensor/Transmitter Time Constants and Sensitivity

Temperature sensors

Besides the measurement dead time resulting from installation limitations described above, significant first-order time lags (in the order of 50 s or more) are created by temperature-sensing elements, espe-

TABLE 2.1   Thermocouple Time Constants

Exposed junction, in water: < 0.25 s
Grounded junction, in water: 1/4-in sheath: 1 s
Ungrounded junction, in water: 1/4-in sheath: 2 s

In air at 65 ft/s for:	Wire or sheath size, in inches					
	1/32	1/16	1/8	3/16	1/4	3/8
Grounded exposed junction wire	1.8 s	4.5 s	10 s	20 s	45 s	85 s
Ungrounded sheath probe	2.7 s	7 s	15 s	30 s	70 s	130 s

SOURCE: *Omega Temperature Handbook*, vols. 25 and 26.

cially if they are installed in thermowells, which introduce additional time lags. A bare thermocouple will give the fastest—practically instantaneous—response, while a filled system bulb in a well will give probably the slowest (Table 2.1).

Flow sensors

Although flow measurement is always technically problematic, the time lags associated with the instruments and installation are small, especially where flow of liquids is involved.

All flow meters are designed to measure the instantaneous velocity of a fluid, and all have a minimum velocity requirement, but only a few can *practically manipulate* that velocity to create a higher useful range. For example, pitot tubes, elbows, vortex, ultrasonic, mass, or target-type flow meters require minimum fluid velocities to generate a meaningful signal and substantial lengths of straight-line runs for accuracy and repetitivity.

The velocity will obviously be determined by such factors as the diameter of a pipeline and the flow. Consequently, to accommodate the installation requirements, a substantial length of the pipeline must be reduced in size to generate a higher velocity and signal range.

Most of the time, it is neither practical nor convenient to reduce a line section if the maximum allowed line size for a particular flow meter is smaller than the rest of the line, which is already determined based on mechanical considerations. This automatically rules out a whole family of flow-metering devices which are otherwise perfectly adequate.

The instruments that *can* adapt themselves best to very low fluid velocities are the ones that can create localized restrictions without distorting the measurement signal, such as orifice plates, flow nozzles, venturi tubes, and positive-displacement and variable-area-type instruments. (See Ref. 1 for other general requirements.)

Pressure sensors

Although pressure measurement is basically simple, and pressure transmitters are rugged, accurate, and fast, some of the largest system first-order time constants are attributable to pressure variations, like those in high-pressure steam boiler load control.

Level sensors

Level sensors can be finnicky, especially if they are of the differential pressure type, which need to be compensated for product density. Their installation requirements and specifications are very important. It is all too common to specify the wrong sensor for the application. This will result in erroneous readings with an unpredictable time lag, which will translate to the rest of the control loop. Otherwise, there are no significant time lags in level measurement.

Analyzer-type sensors

Analyzers, such as those for pH, O_2, redox, dissolved oxygen, and photometric measurements, in general have become much more rugged, and analyzing transmitters are being routinely included in control loops. However, they still require substantial maintenance, and redundant systems are necessary if they are to be heavily relied upon. Installation requirements will unavoidably introduce some dead-time lag.

Electromechanical sensors

Electromechanical devices such as those for sensing revolutions per minute or displacement or for encoding position are in general very reliable and do not need calibration—a major advantage. Accuracy and speed of response are major issues for all mechanical measurements.

Electric power system transducers

These sensors convert alternating-current (ac) variables, like voltage, current, active and reactive power, frequency, and power factor, into linearized direct-current (dc) milliampere signals or into electric impulses. Most types need periodic test-bench calibration, if accuracy and stability are important, as is often the case in frequency and active-power measurements. They also have a considerable response time, which makes them often unsuitable for real-time fault clearing and absolute stability analysis.

Transmitter/Sensor Gain

In most process control applications, a gain factor different than 1 is generated by the range (span) of the transmitter and, to a lesser extent, by the sizing of the control valve. Rarely is a gain factor generated by the process itself. The controller then introduces only the necessary adjustment to reach the dynamically optimum loop gain.

In order to determine the gain of the transmitters in the control loop, data on working conditions is required for most processes, e.g., pressures, maximum flow, and level transmitter range.

Obviously, transmitter data must belong to the particular instrument that is connected to the control loop. This is very important if the proper controller settings are to be obtained.

Since the gain of the controller will be affected by the gain of the transmitter and valve, it is necessary to figure these out before attempting to adjust the controller gain. For example, if a pressure transmitter is calibrated to 1200 psi full output and 700 psi zero output, the transmitter span will be 500 psi (i.e., 1200 psi − 700 psi).

If the system pressure set point is 1100 psi, and the discharge pressure is 100 psi, then the pressure transmitter gain will be

$$G_{Tp} = \frac{\text{set = point pressure − discharge pressure}}{\text{transmitter span}}$$

$$= \frac{1100 - 100}{1200 - 700} = 2$$

The equation is similar for a temperature transmitter. For example, for a system set point of 100°F, a system discharge temperature of 70°F, and a transmitter range of 50 to 150°F, the temperature transmitter gain is

$$G_{Tt} = \frac{100° - 70°}{150° - 50°} = \frac{\text{set-point range}}{\text{transmitter span}}$$

$$= \frac{30°}{100°} = 0.33$$

Transmitter gain is a major controller adjusting factor and can easily range between 0.1 and 10. In these equations, *set-point range* is the difference between the operating state level and the final (deenergized) state level.

Control element gain should not affect loop gain so dramatically, but this gain must be verified and accounted for if different than 1.

Also, the ranges for the controller, transmitter signal, and valve signal must be the same, e.g., 4 to 20 mA or 3 to 15 psi. Some controllers must have their output range limited as part of the adjustments, as for example in antireset wind up control.

Some valve actuators may have smaller ranges. This is the case for control valve split-range designs, where what counts is the combined gain of both valves, working as one from the controller. If a change in type of signal occurs within a loop as, for example, with an electronic/pneumatic transducer, the ranges must all correspond on a 0 to 100 percent scale basis. That is, full-scale and zero-scale correspondence must be maintained overall, otherwise a new gain coefficient is introduced.

With the exception of dead time, linearity is assumed overall. That is, transmitters and control elements have inherently linear responses. PID controllers are linear by definition.

Characterization must be accounted for. All transmitters are assumed to be linearized for any flow, pressure, or temperature condition, as are control elements.

Time-Constant Determination

To determine a time constant requires more ingenuity than work. Here are some general guidelines.

Empirical method

In this method, the response to an input step is measured. The procedure is

1. Recognize the largest time constants to be measured.

2. Design a test that will produce data that can be easily analyzed, for *each* time constant. Second-order response curves are *not* easy to analyze.

3. If the system is under a fluctuating load, stabilize the process either by transferring the load fluctuations to another unit or by arranging a period or interval at constant load with the production people.

4. Set the loop to be measured to manual mode.

5. Introduce a small but significant, say 10 percent, step change in the system. This step must be either a load change (e.g., steam flow) or a control variable step (e.g., fuel).

6. Measure the response of the affected variables. Instruments for this purpose must be accurate and reliable. If response is slow enough, take readings at fixed time intervals and chart them. If response is too fast, a plotter must be used.

7. The time constant is the time it takes for the initial slope to intersect the final value (see Fig. 2.1 response curve). Convert all readings to a percentage of the physical variable by defining a *base* value.

8. This method requires some trial and error until a satisfactory set of readings has been obtained, but has the advantages of not disrupt-

ing the operation too much, requiring only a minimum amount of instruments, and usually taking a short time (approximately 10 min for each test, with 1 or 2 min of actually taking readings).

9. Determine the largest time constants of your system, beginning with the largest and working down. Stop if (a) the next is the controller time constant, (b) you have three already, or (c) the next time constant is less than, say, $\frac{1}{20}$ the largest one (this actually depends on the accuracy required; $\frac{1}{10}$ is sufficient for most practical cases).

Analytical method

This method is recommended when all relevant parameters are known, which is rarely the case. There is one case, though, in which this method is especially useful: level control.

Question What is the integration constant of a tank or drum?

Answer The integration constant is the allowed level fluctuation times the tank surface, divided by the maximum flow.

This is a case similar to the reactor temperature time constant, since we can determine two time constants: one for rising level and one for dropping level. But only one is in the control loop, and that is the one that counts for the controller settings.

The maximum allowed level fluctuation is not necessarily the transmitter range, but is usually a much smaller value, since we are considering normal operating conditions that the level controller maintains before the level alarm goes off. Therefore, level transmitters introduce gain factors smaller than one. This transmitter gain must be compensated by the controller, which adjusts its own gain accordingly.

The largest time constants often come from not the process (i.e., plant), but from the control equipment itself, such as the control valve, with its pneumatic signal amplifier and tubing, or the temperature sensor. This is definitely the case when flow is controlled with the flow meter and control valve on the same pipeline. Such situations require that the time lag of the controller be much lower than the time lag of the transmitter, and the error of the transmitter be much lower than the error of the valve. Otherwise, the control loop is self-defeating. Although widely used, this kind of "self-control" loop should be avoided because it slows down the response of the control system.

"Much lower" or "much higher" is equivalent to a factor of 10 or more in practical process control terms.

On-line system identification

Process control loops can be separated into two parts: the plant and the controller. Since we already know what the controller can do, our major task is to model and identify the plant with the transmitters and control valves. With PID controllers, we can handle only three plant time constants, therefore we specify a plant model composed of gain and the three largest time constants. This makes the identification process substantially easier, although not simple. By measuring at regular intervals the signals from the controller to the plant and reading the plant's response, we can determine the actual values of the time constants. Of course, the plant's response will be contaminated by noise coming from the disturbance input, but this eventually will be eliminated by a *least squares* filtering algorithm, which can easily be implemented in a digital controller. (A well-documented example is in the discussion section of Ref. 2.)

Although an initial gain estimate can be made by the old-fashioned paper-and-pencil method and dead time can be measured by stopwatch, these parameters also can be determined by completely automatic means.

Time-constant simulation

In the end, it may be more productive to simply *create* time constants large enough at the controller input, so that the process time constants can be disregarded when they are too small. PID controllers do not normally have this feature, but it certainly would simplify controller tuning if settings could simply be chosen from a library.[3]

Distributed Control Systems Software[*]

Distributed control systems (DCS) have brought to the control room unlimited possibilities. Initially, we only see the advantages and are slower to recognize the pitfalls. Here are a few of the latter.

Efficiency and flexibility are higher, but so are cost and complexity

Contrary to the expectations we may have, reliability of highly sophisticated control systems is lower than that of old and proven technology. This is because a DCS is basically a collection of computers and is vulnerable as such. Computer-based control systems sometimes have

---

*Based on notes from the course on distributed control systems given by New York Chapter, IEEE, 1985.

the habit of "locking out" any attempts to control the process, and the whole system must be shut down and restarted. With single-loop controller technology, controllers would fail gradually, or only one at a time. This would still allow the plant to function, and an orderly shutdown was easier.

Many plant designers still prefer to play safe and include an additional level of single-loop controllers, or at least hard-wired manual overrides, regardless of the additional expense.

If proper fail-safe procedures are not in place, or have been overlooked, a literal meltdown can occur, since the computer will lock out attempts to control the process as soon as the program "crashes." Hard-wired interlocking is necessary to avoid catastrophic failures of this type.

High operation and maintenance costs

Another unexpected consequence of installing a DCS is its intensive maintenance requirements. Design, operation, and maintenance of DCS require skilled personnel, and expensive parts. Like any computerized equipment, it needs to be constantly "nursed." This means additional personnel, with professional expertise.

Moreover, programs need to be updated. Functions must be added. And we must ask whether the software was correctly designed to deal with any emergency.

Document. Document. Document!

Since configuring a DCS can be so involved, proper documentation must be generated during *all phases*. This documentation need is not always recognized, and this can spell disaster because lack of adequate documentation makes *any* equipment useless. Besides, it takes months to learn how the system works, so *at least* two people must be thoroughly familiar with it, in order not to create a personnel dependency.

Documentation must include detailed flowcharts and loop and logic diagrams. How many times do we ignore at least some of these crucial needs?

Loop diagrams must include "soft" wiring details of program routine calls necessary to locate the specific function at the right address. See Figs. 2.4 and 2.5 for specific examples. Sequential and batch operations must be flowcharted in as much detail as possible. Start and stop interlocks for pumps, etc., can best be described by logic diagrams (see Appendix).

Extensive design with ladder diagrams is not recommended because

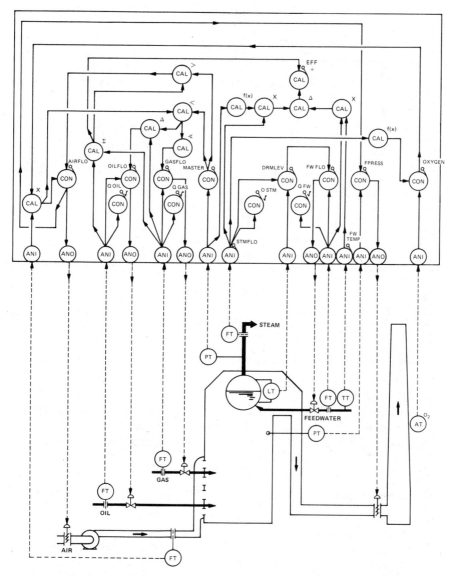

Figure 2.4 Configuration diagram for boiler control system in Fig. 2.5: ANI: input, ANO: output, CAL: calculation, and CON: control or integrate. (SOURCE: Fisher and Porter Company)

the potential for mistakes is large and the diagrams are not easy to interpret. At some point, the logic may be translated into wiring diagrams that require a ladder, or "elementary" format, but these are based on a logic diagram that helps to keep the functional aspects clear.

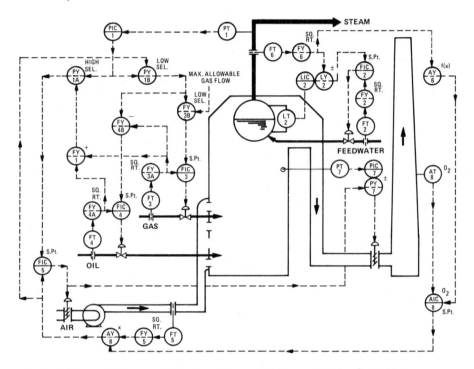

Figure 2.5   Typical boiler control system. (SOURCE: Fisher and Porter Company)

All this may increase engineering expenses substantially but will pay off many times over and will generate the admiration, respect, and gratitude of those who are responsible for keeping the system running.

Explaining programming code only is absolutely not sufficient, by a long shot.

input/output requirements

There are two basic type of signals: analog and discrete. There are two basic type of functions: measuring and interlocking. An analog signal is one that changes smoothly; it carries a certain amount of error. A discrete signal, on the other hand, is formed by a series of switches that can either *latch* or send out a *pulse*. The pulses represent a code that has to be interpreted accordingly. Both measuring and interlocking functions can be realized from discrete or analog signals.

Measuring function requirements

The largest limitations in accuracy come always from the field measuring devices, i.e., transmitters, transducers, and signal sensors and converters. The DCS input will have its own accuracy limitations, but these can be lowered to the point where they will not significantly add to the field device error. A good criterion is that the *DCS input will be 5 to 10 times more accurate than the measuring device.*

In addition, for discrete signals, timing is crucial. The pulses not only must have a maximum frequency, but also a minimum duration in order to be picked up by the system's input card. Pulse counting should be done at the input card level, in order not to tie up central processing unit (CPU) time.

Pulse signals can be two- or three-wire. Three-wire signals provide a complementary verification signal, to avoid loosing pulses. Binary-coded decimal (BCD) signals require multiple discrete inputs. Latching by either the field instrument or the input card is required in order to pick up the signal.

An 8-bit digital signal provides roughly 0.4 percent accuracy, but nothing is lost at the input level if the field sensor is equally accurate or better.

The advantage of using BCD signals, and discrete signals in general, is that there is no drift error from an amplifier output stage, as would be the case for an analog signal of, say, volts or milliamperes.

Table 2.2 compares accuracies and bits.

Industrial instrumentation accuracy is a long-debated subject. To assume anything better than 1 percent for an installed device under actual operating conditions is being optimistic. Most process control transmitters can be assumed to have an installed accuracy around 2 to 5 percent. In some cases, e.g., for pressure and temperature measurements, this can be easily verified, since it is easy to install side-by-side

TABLE 2.2   Relations of Bits, Accuracy, and Significant Digits

Bits	Accuracy		Significant decimal digits
1	$\frac{1}{2}$	or 50%	1
2	$\frac{1}{4}$	or 25%	1
3	$\frac{1}{8}$	or 12.5%	1
4	$\frac{1}{16}$	or 6%	2
5	$\frac{1}{32}$	or 3%	2
6	$\frac{1}{64}$	or 1.6%	2
7	$\frac{1}{128}$	or 0.8%	3
8	$\frac{1}{256}$	or 0.4%	3
9	$\frac{1}{512}$	or 0.2%	3
10	$\frac{1}{1024}$	or 0.1%	3
11	$\frac{1}{2048}$	or 0.05%	4
12	$\frac{1}{4096}$	or 0.025%	4
13	$\frac{1}{8192}$	or 0.012%	4
14	$\frac{1}{16,384}$	or 0.006%	5
15	$\frac{1}{32,768}$	or 0.003%	5

accurate thermometers or pressure gauges. This is not the case for flow meters or pH meters, where complicated steps must be taken in order to verify the readings. Required accuracy must be stated for all measurements, and, if necessary, verified in the field with the instrument installed.

Special consideration must be given to those measurements that require very high accuracies, say within less than 1 percent. In those cases, accuracy has to be separated into its component parts, such as hysteresis, resolution, linearity, repeatability, temperature coefficient, and drift to determine exactly which ones are going to significantly affect process measurement.

There is therefore no point in stretching the input card accuracy to more than 10 times the field instrument reading accuracy, since this will not significantly affect the field reading. Also there is no point in displaying more significant digits than the accuracy of the field device requires; usually three significant digits is enough. For totalizers, only the necessary significant digits *between consecutive readings* need to be displayed. *It is meaningless—and confusing—to show more significant digits than those necessary to cover field instrument accuracy.* Anything beyond the significant digits should be trailing zeros.

Scanning rates of digital input devices should be not less than 5 times the smallest time interval to be sampled.

Fail-safe interlocking requirements

Interlocks protect the equipment and avoid hazardous conditions. All interlock functions and alarms must be fail-safe; that is, if the interlock function fails, it takes a safe action, usually an orderly equipment shutdown and sounding an alarm. Generally,

- Equipment shuts down if a control loop loses power or continuity. Interposing relays must fail "open"; they must maintain or assume an open position at failure.

- An alarm goes off if an input circuit loses power or continuity. Interposing relays must fail in the closed position.

Outputs in general

The same considerations apply to controller outputs as to inputs, but with minor differences. Each analog output that comes from one controller is counted as a loop.

Analog outputs must be fail safe also. There are four possibilities for failing safely: (1) going to zero, (2) going to maximum value, (3) freezing in position, and (4) assuming a predetermined position. The choice depends on the final control element.

Discrete control outputs must fail in the open position, to deenergize the equipment. Alarm outputs must fail in the closed position, to energize the corresponding alarm device. Interposing electromechanical relays adds a fraction of a second—approximately 0.1 s—to any output execution.

Fail, but Fail Safely

"People make mistakes. Machines break down."

There is no question that sooner or later, our installation, system, machine, structure, integrated circuit, alarm, safety device, or whatever, is going to *fail*! We have to be ready for this moment, and not delude ourselves with excuses like: "It's a remote possibility," "Improbable," "one chance in a billion," "The operator will pick it up," and so on. The process control engineer and designer must be particularly sensitive about this subject, because their responsibility is to suggest and design fail-safe configurations.

What is a fail-safe system? Obviously a system that, when control is lost, will do the least amount of damage—an amount that is both tolerable and *expected*. Let's look at some examples of fail-safe and not-fail-safe systems.

As we all know from our daily experience, a car on the highway is a not-fail-safe system. There are too many factors, human and mechanical, that can go wrong and create an accident condition: a slippery road, a brusque maneuver, momentary blindness, confusion. Actually, one may wonder why there are not many, many more car accidents, considering all the possibilities for mistakes. Cars should be equipped with more interlocks, velocity-distance detectors that automatically keep a safe clearance, and with computers that maintain safe driving conditions during maneuvering.

Another case of poor fail-safe design is manually operated railroad signal systems. These are very good illustrations of how unreliable human beings are, and to recognize this is the first step to fail-safe design.

The ultimate example of not-fail-safe design is the present generation of nuclear reactors, because, when all control is lost on a nuclear reactor, the worst possible scenario becomes a reality: extensive loss of life and property damage for extended periods of time.

During design, construction, start-up, operation, and maintenance, the worst-case scenario, or catastrophic failure of each component/system, should be kept in mind as the *first* concern. Sometimes, there is no way to avoid a possible worst-case scenario, regardless of all the backups and redundancies. This is what is known as being "in for the ride."

There are a few, very simple rules for designing for fail-safety:

1. *Provide backup:* For example, if a diesel generator system has, say, a 10 percent or 1 out of 10 chance of tripping or failing to start, a backup will reduce the chance of simultaneous failure to 10 percent of 10 percent—1 in 100. If the resulting mean time to failure exceeds the life expectancy of the installation, we are on the right track. Backup is an attractive way to limit economic losses from equipment downtime or unavailability, and backup design is especially attractive for controls because of their relatively low cost. It is not always feasible, however; for example, it would be economically impossible to back up a nuclear reactor.

2. *Design for intrinsic fail-safety:* If an intrinsically fail-safe system fails, there will not be immediate consequences like loss of life or property damage. For example, if a pump impeller snaps, the housing will contain the damage.

3. *Include plenty of safety devices:* Relief valves, fuses, circuit breakers, and so on are relatively low-cost devices that normally protect huge investments and human life. The importance of adequately designed, installed, and maintained "safeties" cannot be emphasized enough.

4. *Use interlocks:* Certain conditions must be met to start up, shut down, or run equipment. If these conditions are not met, an interlock must be engaged, usually to avoid human error. Interlocks must be designed to be fail-safe also, so that if the interlock fails, the system/installation will shut down safely. Manual interlock overrides absolutely cannot be tolerated during normal operation.

5. *Install alarms:* The only purpose of alarms is to alert the human operator of impending—short or long term—trouble, so that the operator can take remedial action beyond the scope of the automatic controls, such as trying to bring additional equipment on-line manually or scheduling maintenance. An alarm should always operate in a fail-safe mode so that if it fails, for example, by an open circuit, the alarm will be turned on—by a back up alarm, if necesary.

6. *Build in tolerance:* Some equipment has a very high degree of reliability. An example is a properly installed low-pressure water pipeline. The worst that can possibly happen is that it can rust through or break because of structural support failure, but if proper cathodic protection and drains are installed, no equipment or life will be at risk. In these cases, to design in redundancy is useless because the redundant elements will deteriorate at the same rate! The risk of failure is so remote and the consequences are so insignificant that the system ultimately becomes a calculated and affordable risk. Nevertheless, all failure scenarios must be analyzed and accounted for.

7. *Ensure proper maintenance:* Preventive and predictive maintenance are the tools to reduce untimely failure and its consequent
plant and system shutdowns. Adequate upgrades, records, and re-
ports are the key to successful maintenance.

We are aware that systems and devices are not fail-proof by any
means, but we are usually able to forecast how they are going to fail to
a large extent. From the previous analysis, we realize that the probability of failure must be weighed together with the *consequences* of
failure.

Continuous In-Plant Monitoring of Compressor Performance and Plant Air/Gas Consumption, through Pressure Measurement Alone*

How can a plant continuously monitor compressor performance?
When is a compressor due for overhaul? How can plant air/gas load be
measured? How can compressed air/gas leaks be accounted for?

By using the law of perfect gases, and pressure measurement alone,
there is a simple way to measure *continuously* air/gas flow coming
from reciprocating and positive-displacement compressors, as well as
air/gas consumption.

First determine the volume of the system, up to the pressure reduction valves. This includes the receivers, accumulators, and piping.

Next, measure the time interval it takes for the pressure to change
over a certain range P_1–P_2 when (*a*) the compressor is running and (*b*)
when the compressor stops. The flow is

$$Q_{air} = \frac{\text{system volume} \times (P_1 - P_2)}{\text{time interval} \times \text{atmospheric pressure}}$$

The first reading, *a*, will tell the amount of excess air being pumped
into the system. The second reading, *b*, will tell the amount of air being used. The sum of *a* plus *b* tells the amount of gas being produced
by the compressor:

$$c = a + b$$

There are different ways to implement this monitoring loop. Here
are a few:

1. Use the start/stop signals from the compressor to take pressure
readings.

---

*Adapted from Ref. 4.

2. Use a calculation-block function from an electronic controller. Use a logic algorithm to determine what flow you are reading: When the pressure increases the compressor should be pumping, and when the pressure decreases the compressor should be unloaded.

3. Use any combination of the above.

These "readings" can be compared to threshold or nameplate values and an alarm can be activated indicating *compressor overhaul due* or *gas leak probable*.

The accuracy of this indirect measurement will depend on (1) the constancy of the air/gas load, (2) how "perfect" the gas is at that pressure and temperature, and (3) the accuracy of the instrumentation. The ambient temperature may play an important role if high accuracy is required, and temperature readings in different physical locations may be necessary, but the compressor discharge/average ambient temperature ratio will provide sufficient correction.

Gas compressor pressure control has many similarities to level control. The system's pressure time constant will be

$$T_{\text{compressor system}} = \frac{\Delta p}{p_o} \frac{V_{\text{system}}}{Q_{\text{gas max.}}}$$

where Δp = allowed pressure change
p_o = discharge pressure
$Q_{\text{gas max.}}$ = maximum compressor output
V_{system} = volume of receivers and piping, up to pressure reducing valves

References

1. Flowmeter Selection Chart, Fischer and Porter Company, Hatboro, Pa.
2. R. E. Kalman, "Design of a Self-Optimizing Control System," *Transactions of ASME*, vol. 80, February 1958, pp. 468–478.
3. M. Polonyi, "PID Controller Tuning Using Standard Form Optimization," *Control Engineering*, March 1989, pp. 102–106.
4. M. Polonyi, "Operating Performance of Reciprocating or Positive Displacement Compressors," *Chemical Engineering*, Dec. 9/23, 1985, p. 132.

3

Standard-Form
PID Controller Tuning

By applying the Graham and Lathrop–type standard forms to the characteristic equation of a process control loop containing a PID controller, a solution can be obtained for each of the adjustable parameters of the controller, i.e., gain, integral action, and derivative action. These solutions, together with the transient response plots derived from them, allow exact tuning of PID-type controllers once the time constants of the plant have been determined. This eliminates any uncertainty about the response. There is no need for computerized optimization or simulation once a specific response curve has been chosen. The process is represented by first-order time constants, with or without dead time. To obtain an exact solution, proportional, integral (PI) controllers need the two largest time constants, while PID controllers need the three largest.

Besides describing standard-form optimization, this chapter analyzes the effects of derived PID controller settings for widely differing time constants.

An interesting result of the standard form method of synthesis is that it shows the need for a large time constant upstream of the PID controller in order to lower control element power supply requirements. This time constant should come either from the process itself or should be incorporated into the controller as an option.

Background

PID controller tuning has remained an obscure art rather than an exact science. The reasons for this are simple: nobody really knows what the settings should be, since all criteria are qualitative in nature and, because of the large self-regulatory capacity of most process systems, the margin of tolerance is high. Accurate settings are therefore not necessary.

A fast response is always desirable. However, in practice the results may not be good; because of the unknown nature of many parameters, the system may become unstable or stop controlling altogether. A slow response may then be necessary. Another good reason to choose a slow response is that it is more efficient in energy terms.

The tuning of process controllers is an activity as old as process control itself. We know by now that Watt's flying-weight regulator embodies what we call gain (or proportional band) in present process control technology; i.e., it is a negative feedback amplifier.

The question comes to mind again: What are the best settings for these proportional, integral, derivative process controllers. There is still no simple answer. As a consequence, we may state as a fact that the great majority of industrial controllers are poorly tuned.

Accurate settings are not that important because the plant/process has a lot of built-in tolerance or time lags. For most practical process control applications, accuracy within 10 percent is sufficient. Problems become obvious only during large load upsets, which are rare. The PID tuning problem also has nothing to do with steady-state response. Transient response is important, but may follow vastly different routes to arrive at the same end, as long as it does so within a prescribed error baud and time.

In 1953, Graham and Lathrop[1] published a paper listing different normalized transient responses, based primarily on results from the integral of the time of absolute error (ITAE) criterion. But at the same time, the paper highlighted the elusive nature of the tuning problem.

By using standard-form optimization and Graham and Lathrop's list, a set of algebraic equations can be derived for each PID controller (see Refs. 1 and 2 for a list of standard forms). The equations can also be used for a self-tuning controller, by incorporating a time-constant identification algorithm in the controller hardware. See Fig. 3.1 for a sample of PI controller responses.

Standard-Form PID Tuning

The features of the standard-form optimization method are[3]:

1. It offers a broad choice of responses: fast and slow or energy-efficient (ITAE, Butterworth, binomial, zero-displacement error, etc.).

2. It eliminates the uncertainty which comes with the use of empirical methods, which are effective only for very simple control loops.

3. It avoids making assumptions that do not necessarily occur, such as

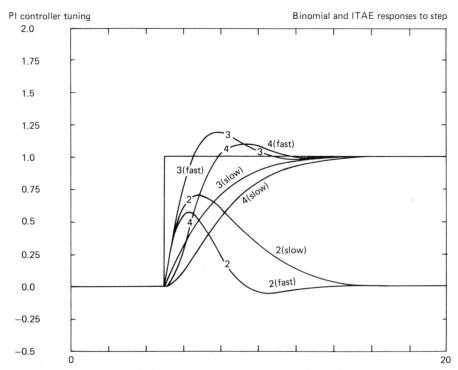

PI controller tuning   Binomial and ITAE responses to step

Figure 3.1   Example of responses using a PI controller and standard-form tuning. Time constants: $T_1 = T_2 = 1$; controller settings: binomial: $G = .33$, $T_i = 1.8$ (slow response); ITAE: $G = 1$, $T_i = 1$ (fast response). Curves: 2: regulator output; 3: controller output (after T_1); 4: servo output.

those about the number of time constants and relations between the spectral and transient responses.

4. It provides a set of algebraic equations. The resulting settings are dynamically exact.

5. It requires no computer simulation, iteration, or analysis. It is therefore unique and easy to use.

6. As noted, a PID controller needs three time constants; a PI controller needs two.

7. Transient response curves are readily available.

8. It requires only approximate knowledge of the system's largest time constants.

The disadvantages are:

1. Only a limited number of time constants can be accommodated.

2. Large time constants must be added if the process does not provide them naturally.

3. It does not take into account the control effort to meet the criterion.

4. Solutions involving dead time are cumbersome and of limited use.

5. It does not provide information on controller output overshoot.

6. It requires dynamic simulation to verify the response for control effort overshoot and dead-time stability.

Plant Models

Most nonelectrical (heat-transfer, mechanical, chemical, or fluid-dynamics) open-loop processes can be characterized by a series of first-order time constants, with unity gain:

$$\text{Plant} = \frac{1}{(1 + sT_1)(1 + sT_2)\cdots}$$

In addition, process control loops have dead time, and their gains are different from one. In most cases, gain and dead time do not originate in the process itself, but in the way we monitor and control it. Most dead time is caused by our inhability to monitor a particular process at the source, or while it is developing, and we have to content ourselves with a measurement that produces dead time. In the same way, gain is created when we choose the range of the signal transmitter that is put into the control loop. This applies to the control element too, especially if we need a large control effort in order to keep setpoint excursions small.

These considerations are relevant to the solution of our problem. First we will establish a general model.

Input and Output Definition

For any control loop, we define three outputs for each input. We call them the *servo output* (Fig. 3.2), the *regulator output* (Fig. 3.3), and the *controller output* (Fig. 3.4). For servo and controller outputs,

$$\lim_{t \to \infty} \frac{c(t)}{i(t)} = 1$$

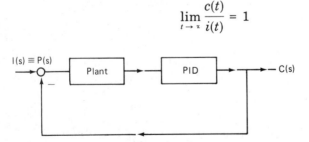

Figure 3.2 "Pure servo" control loop block diagram.

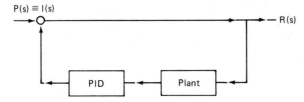

Figure 3.3 "Pure regulator" block diagram.

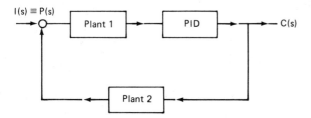

Figure 3.4 Controller output block diagram.

where $c(t)$ = control output and $i(t)$ = set-point change. For regulator output,

$$\lim_{t \to \infty} \frac{r(t)}{d(t)} = 0$$

where $i(t) = d(t)$, since $d(t)$, the input to a regulator, is also called a *disturbance; r(t)* = regulated output.

For the "pure servo" control loop in Fig. 3.2, the functional transform is

$$\frac{C(s)}{I(s)} = \frac{\text{plant} \times \text{PID}}{1 + \text{plant} \times \text{PID}}$$

For the "pure regulator" loop in Fig. 3.3,

$$\frac{R(s)}{P(s)} = \frac{1}{1 + \text{plant} \times \text{PID}}$$

For the "controller" loop in Fig. 3.4,

$$\frac{C(s)}{P(s)} = \frac{\text{plant1} \times \text{PID}}{1 + \text{plant1} \times \text{plant2} \times \text{PID}}$$

These definitions illustrate the fact that *all control loops are regulators and servos at the same time.*

The controller output, which is a servo-type output, must be checked for "reasonable" response, otherwise unrealistic demands can be imposed inadvertently. The larger the available control effort, the faster the response can be, but this depends on the available control power source as well. If unrealistic control efforts are required during a transient to meet a criterion—say, twice the steady-state requirement—a low stability margin is indicated. In practical applications these loops will not be able to function as predicted since smaller time lags and nonlinearities will now become evident.

Generalized Block Diagram Example

Figure 3.5 is a block diagram representation of a high-pressure steam boiler load master control loop. Note the following facts:

1. An input and an output can be defined before and after each block.

2. Steam demand can be measured directly with a flow meter.

3. Steam production cannot be measured directly with a flow meter on a stand-alone boiler (see Chap. 9)..

4. The integral of the *instantaneous differences* between steam demand and steam production translates into steam pressure swings.

5. Steam demand is a signal input but a process output! Consequently, it is a negative input.

6. Each block must be represented at least by a gain > 0.

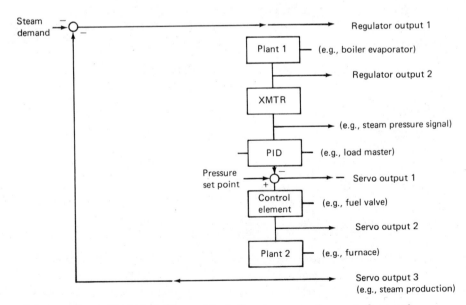

Figure 3.5 Generalized block diagram for boiler steam-pressure control example.

Solutions to the Characteristic Equation for PID Controllers

We know from standard form optimization that an ideal PI controller can handle (and needs) two process time constants. If the larger one, T, is an integration time constant, then the solution will be[2]
Gain:

$$G = \frac{bT}{a^2 T'}$$

Integral action:

$$T_i = abT'$$

Natural frequency:

$$\omega = \frac{1}{aT'}$$

where a and b are coefficients from the standard form of choice and T' is the smaller first-order time constant. However, it is always necessary to verify that the controller and control element can respond to the demands imposed if large gains are involved. See Fig. 3.6.

If the largest process time constant (not dead time) precedes the input to the controller, a filtering and smoothing effect takes place, which makes high gain settings feasible. This time constant can also be artificially generated by software or hardware. This would be Plant1 in our generalized block diagram, Fig. 3.5. If the largest time constant does not precede the controller input, saturation may take place and the response will be totally different. See Fig. 3.7.

Also, the nature of the process must be considered before we can make a decision on how strong our control action can be. It can be almost free, as in controlling the power factor of an alternator with the excitation field. So the question arises "What should the settings be for a smooth controller response?" For smooth response, a second-largest time constant may be created, equal to the main time constant. This yields the lowest possible settings,[4] and an equally moderate response from the controller. See Fig. 3.1.

Natural Frequency

Observe from the equations that the natural frequency of the closed-loop system, a measure of the velocity of response, is determined by the smallest time constant, not the largest as would be reasonably expected. Consequently, only the largest time constants—all of the same order of magnitude—need to be included. If there is only one large

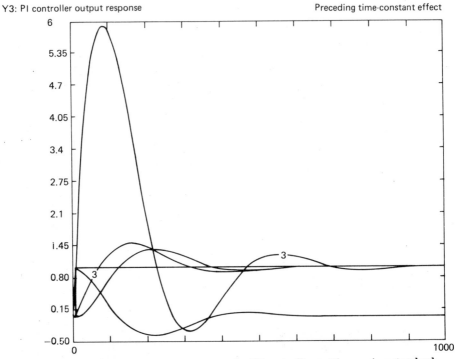

Y3: PI controller output response   Preceding time-constant effect

Figure 3.6   Zero-displacement ITAE response for PI controller settings using standard-form tuning, with process time constants 50 and 500. Gain = 7.5; T_i = 150. The servo and regulator responses are the same, but observe the tremendous overshoot when the controller is preceded by the smaller time constant (50) only.

time constant, and including the second one creates an unacceptable burden on the controller output, then it should be replaced, as just described, by a *larger time-constant value, generated at the controller input.*

Speed and magnitude of response can still be influenced by choosing a criterion that will yield an acceptable response. This is another good reason why settings are not critical. Therefore, a lookup table can be developed that takes care of most practical situations (Table 3.1)

Dead Time

To find solutions for loops with dead time, it is necessary to use an approximation of the Laplace representation of dead time e^{-sT}, such as $(1 - sT)$, or $(1 - sT/2)/(1 + sT/2)$ (see dead-time response examples based on standard-form optimization of PID controller settings in Figs. 3.8 to 3.16). For dead-time approximations, using deriva-

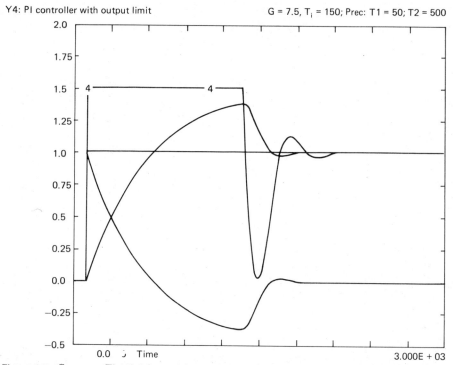

Y4: PI controller with output limit   G = 7.5, T$_i$ = 150; Prec: T1 = 50; T2 = 500

Figure 3.7   Same as Fig. 3.6 but with saturation. Curve 4 is the control element response. Preceding time constant: T1 = 50.

TABLE 3.1   Recommended Settings for Systems with Two Time Constants

System time constants	Critically damped or binomial		ITAE zero displacement	
	G	T_i	G	T_i
$T_1 = T_2 = T$	0.33	$4.5T$	1.8	$1.2T$
$T_1 = 10T_2$	3.3	$9T_2$	7.5	$3.7T_2$

tive action simplifies the solution somewhat. This suggests that derivative action may have a beneficial effect in countering dead time in general. On the other hand, the combination of dead time and derivative action imposes again a considerable burden on the controller/control element, which may be unable to respond and will, instead, saturate.

Saturation may not be a major problem in some cases, since it only will slow down the response slightly, as can be observed from simulation curves (Fig. 3.7), while in other cases it will make it outright unstable.

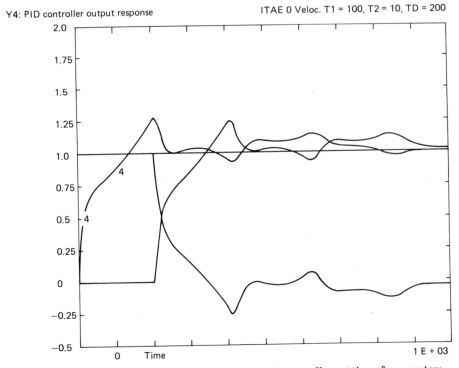

Figure 3.8 ITAE zero velocity standard-form derived controller settings for a system with dead time: $G = 0.7$, $T_i = 144$, $T_v = 100$.

Derivative Action

Although dynamically exact solutions are possible with pure derivative action, a real controller will respond differently:

$$\text{PID} = K_c\left(\frac{1}{sT_i} + \frac{1 + sT_v}{1 + \alpha sT_v}\right)$$

where α is the derivative action *sharpness*. Since most modern PID controllers carry a microprocessor, it is just a matter of the controller squeezing in enough calculations to produce an almost pure derivative action effect. Figure 3.17 shows the effect of derivative action sharpness α.

Figure 3.17 shows a case in which derivative action has a clear, stabilizing, and well-behaved effect. The controller output was simulated with two of the three equal time constants upstream.

For all practical purposes, $\alpha = 0.1$ will provide satisfactory deriva-

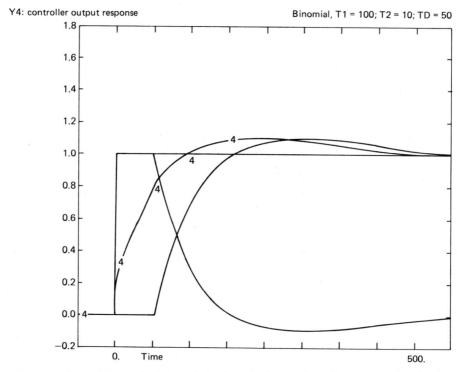

Figure 3.9 Binomial standard-form derived controller settings for system with dead time: $G = 1.2$, $T_i = 86$, $T_v = 25$.

tive action. This requires sampling the input signal ten times faster than would be otherwise necessary.

Developing the Normalized Characteristic Equation

Regardless what input/output pair we choose, the same characteristic equation will result. For any block diagram, such as that in Fig. 3.18, the transfer function is

$$\frac{C(s)}{I(s)} = \frac{G(s)}{1 + G(s)H(s)}$$

and the characteristic equation is

$$CE = 1 + G(s)H(s)$$

Y4: controller output response   Binomial, T1 = 100; T2 = 10; TD = 25

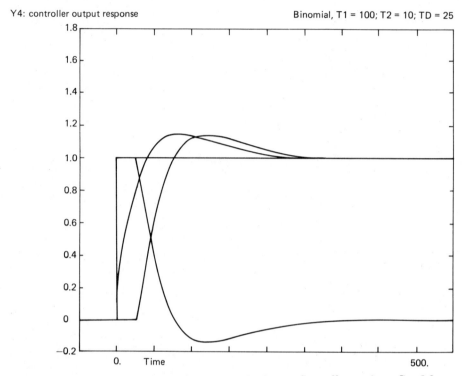

Figure 3.10 Same as Fig. 3.9 but with a smaller dead time. Controller settings: $G = 2.3$, $T_i = 77$, $T_v = 13$.

For a regulator I/O pair, $G(s) = 1$. For a pure servo I/O pair, $H(s) = 1$. Nevertheless, *the characteristic equation will not change.*

In order to use the characteristic equation, we must normalize it. To normalize, we must divide it by the coefficient of the n highest-order term of the s operator:

$$CE = (T_1 T_2 \cdots T_n)s^n + \cdots + As^0$$

The normalized CE is

$$s^n + \cdots + \frac{A}{T_1 T_2 \cdots T_n}s^0$$

Now we are free to equate the normalized CE to a standard form of the type

$$s^n + a\omega s^{n-1} + \cdots + \omega^n$$

where the coefficients are

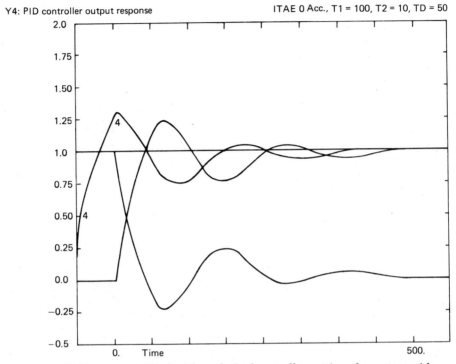

Figure 3.11 ITAE zero acc. standard-form derived controller settings for system with dead time: $G = 2.12$, $T_i = 171$, $T_v = 25$.

$$a\omega = \cdots$$

$$b\omega^2 = \cdots$$

$$\cdots\cdots\cdots$$

$$\omega^n = \frac{A}{T_1 T_2 \cdots T_n}$$

As before, ω is the natural frequency.

When dead time is present there is only a limited solution, or no solution at all. In any event, obtaining a solution is a cumbersome process.

In general, this method yields high gain settings if the largest time constants—also known as *roots* or *poles*—are too far apart, because the natural frequency is affected by the smallest time constant. This is a characteristic feature of this method and comes from trying to fit the response to the criterion. Consequently, there is no point in including time constants that are, say, two orders of magnitude smaller for PID controller tuning.

The controller will produce a large overshoot if the largest time constants are downstream, i.e., after the controller. This is not a desirable

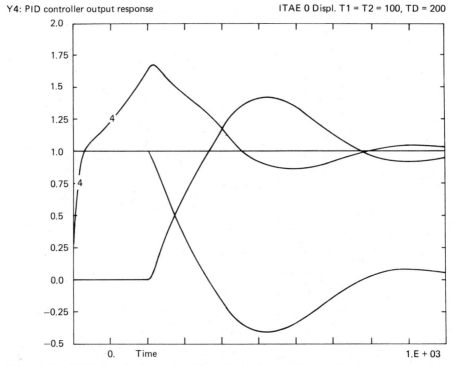

Y4: PID controller output response   ITAE 0 Displ. T1 = T2 = 100, TD = 200

Figure 3.12   ITAE zero displ. standard-form-derived controller settings for system with dead time: $G = 1.03$, $T_i = 196$, $T_v = 100$. Not shown: less overshoot (1.5) with ITAE zero-velocity settings, but more oscillation at the end; overdamped with ITAE zero-acceleration settings.

situation, and it indicates a control loop that will tend to be unstable and slow to respond, *regardless* of the settings and the criterion or the method used to tune the controller.

Solutions can also be found if there are complex roots, although this is rare in power and process control applications.

Exact Solutions for Plants with Integration Time Constant

The following are the solutions for systems with an integration constant T, and first-order T_1 and T_2.

PI controller without dead time
(T: integration constant)

$$G = \frac{bT}{a^2 T_1}$$

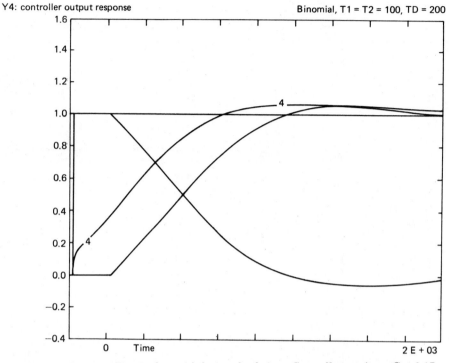

Figure 3.13   Same as Fig. 3.9 but with larger dead time. Controller settings: $G = 0.17$, $T_i = 97$, $T_v = 100$.

$$T_i = abT$$

$$T_v = 0$$

$$\omega = \frac{1}{aT_1}$$

PID controller without dead time
(T: integration constant)

$$G = \frac{cT(T_1 + T_2)^3}{a^3(T_1 T_2)^2}$$

$$T_v = \frac{[b(T_1 + T)_2)^2 - T_1 T_2 a^2]aT_1 T_2}{c(T_1 + T_2)^3}$$

$$T_i = \frac{acT_2 T_1}{T_1 + T_2}$$

Y4: PID controller output   Binomial response, T1 = T2 = TD = 100

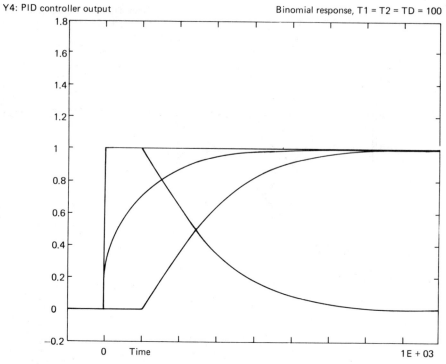

Figure 3.15 Same as Fig. 3.13 but with all time constants equal. Controller settings: $G = 0.7$, $T_i = 201$, $T_v = 50$.

$$\omega = \frac{T_1 + T_2}{a T_1 T_2}$$

PID controller with dead time $2T_d$

$$G = \frac{T}{T_d}\left[1 - a\left(\frac{T_1}{bT_d}\right)^{\frac{1}{2}} + \frac{T_1}{T_2}\right]$$

$$T_i = c(bT_d T_1)^{\frac{1}{2}} + T_d$$

$$\omega = \frac{1}{(bT_d T_1)^{\frac{1}{2}}}$$

$$T_v = T_d$$

Y4: PID controller output response   Binomial, T1 = T2 = 100, TD = 200

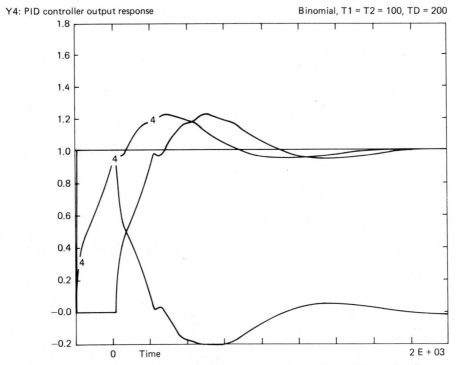

Figure 3.16 Same as Fig. 3.14 but with dead time very large: $T_d = 10\ T_1$ or T_2. Controller settings: $G = 0.05$, $T_i = 68$, $T_v = 500$.

Exact Solutions for Plants with First-Order Time Constants T_1, T_2, T_3, or T_d

PID controller with dead time

$$\omega = \sqrt{\frac{T_d + T_1 + T_2}{b\ T_d T_1 T_2}}$$

$$K = \left[T_d T_1 + T_d T_2 + T_1 T_2 - (T_d + T_1 + T_2)\frac{a}{b\omega} \right]\frac{1}{T_d^2}$$

$$T_i = \left(\frac{c}{\omega} + T_d\right)\frac{K}{1 + K}$$

$$T_v = T_d$$

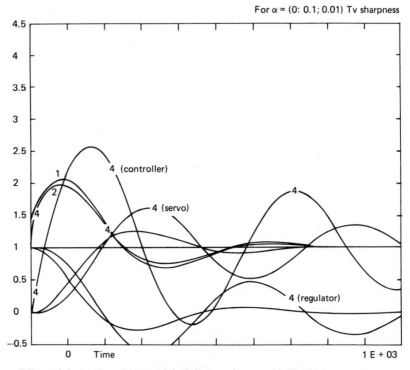

Figure 3.17 Effect of derivative action, with different degrees of sharpness: $\alpha = 0$: no derivative action; underdamped response (curve 4). $\alpha = 0.1$: typical ITAE response (curve 1). $\alpha = 0.01$: some improvement over $\alpha = 0.1$ (curve 2). Plant constants: $T_1 = T_2 = T_3 = 100$. Standard-form-derived controller settings: $G = 2.9$, $T_i = 203$, $T_v = 50$.

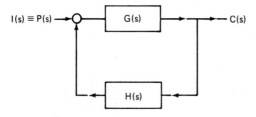

Figure 3.18 Block diagram.

$$\text{Dead time} = 2T_d$$

PID controller with three time constants

$$\omega = \frac{1}{a}\left(\frac{1}{T_1} + \frac{1}{T_2} + \frac{1}{T_3}\right)$$

$$G = c\omega^3 T_1 T_2 T_3 - 1$$

$$T_i = \frac{c}{\omega}\left(\frac{K}{1 + K}\right)$$

$$T_v = \frac{b\omega^2 T_1 T_2 T_3 - (T_1 + T_2 + T_3)}{K}$$

Exact Solutions for Plants with Time Constants, T_1 and T_2

PI controller without dead time

$$\omega = \frac{1}{a}\left(\frac{1}{T_1} + \frac{1}{T_2}\right)$$

$$K = \frac{b(T_1 + T_2)^2}{a^2 T_1 T_2} - 1 = b\omega^2 T_1 T_2 - 1$$

$$T_i = \frac{b}{\omega} - \frac{1}{T_1 T_2 \omega^3} = \frac{K}{T_1 T_2 \omega^3}$$

Symbols, Relations, and Assumptions

s = Laplace operator, or first derivative
G = loop gain
T_i = controller integral action time
ω = loop natural frequency
T = process integration constant
T_2 = process second time constant, first-order: $1/(1 + sT_2)$
$2T_d$ = loop dead time: $(1 - sT_d)/(1 + sT_d)$
T_1 = process first time constant, first-order: $1/(1 + sT_1)$
T_i = controller integral action time
T_v = controller derivative action time
K_t = transmitter gain
K_e = control element gain
K_c = controller gain
K_v = rate gain
K_i = reset gain
p.u. = per unit

Table 3.2 lists standard-form optimization constants.

TABLE 3.2   Standard Form Optimization Coefficients[1,2]

Standard form	PI		PID		
	a	b	a	b	c
Binomial or critically damped	3	3	4	6	4
Butterworth	2	2	2.6	3.4	2.6
ITAE zero error displacement	1.75	2.15	2.1	3.4	2.7
ITAE zero velocity error	1.75	3.25	2.41	4.93	5.14
ITAE zero acceleration error	2.97	4.94	3.71	7.88	5.93
Solution time steady-state error	1.55	2.1	1.6	3.15	2.45

For all controllers

$$K_c = \frac{G}{K_e K_t}$$

$$\text{PID} = K_c \left(1 + \frac{1}{sT_i} + sT_v \right)$$

$$= K_c + \frac{K_i}{S} + K_v s$$

Since the gain of the controller will be affected by the gain of the transmitter and valve, it is necessary to figure these out before attempting to adjust the controller gain. Refer to Chap. 2 for suggestions.

With the exception of dead time, linearity is assumed overall. That is, transmitters, sensors, and control elements have inherently linear responses. PID controllers are linear by definition. All control elements are assumed to be characterized for any flow, pressure, or temperature condition.

Conclusions

The standard-form PID controller tuning method will provide

- Settings for PID controllers once a response curve has been chosen, thereby eliminating the uncertainty and long computations that are necessary otherwise
- Approximate settings sufficient for most power and process control applications

On the other hand, it requires

- At least approximate knowledge of the largest time constants. This information may not be readily available or simple to determine.

- Verification of control effort margins when high loop gains are involved.

- Determination of transmitter and control element gains

References

1. D. Graham and R. C. Lathrop, "The Synthesis of 'Optimum' Transient Response: Criteria and Standard Forms," *Transactions of AIEE*, vol. 72, pt. II, November 1953, p. 281.
2. J. J. D'Azzo and C. Houpis, *Linear Control System Analysis and Design: Conventional and Modern*, chap. 16, McGraw-Hill, New York, 1981, pp. 549–552.
3. M. Polonyi, "PID Controller Tuning Using Standard Form Optimization," *Control Engineering*, March 1989, pp. 103–106.
4. Software package, "PID Controller Tuning—Settings Software," available from STRIDIRON Services, Maspeth, N.Y.

Dynamic Energy Control and Optimization

Boiler Steam Pressure Control

Boiler performance is usually evaluated for steady-state operation. This is hardly the actual operating mode in most applications. Usually, load changes are overlooked because transient behavior is not well-understood. The consequences are a higher fuel consumption and poorer response to system upsets if controller settings are not carefully determined.

Since the fuel/air mixture will never be optimal under varying burner flow conditions, changes in the firing rate should be kept as slow as possible by fully exploiting built-in steam storage capacity. In addition, the furnace heat-transfer time lag must be taken into account, since it plays an important part in finding the "best" controller settings. Feedforward load control must be used with care; it may cause unnecessary fuel valve swings, with subsequent energy losses, if the control signal is not properly delayed.[1]

A dynamic boiler model will be proposed in this chapter, and various control arrangements will be considered. Also, the transient losses of control in general are analyzed.

Furnace time constant

When fuel is burned to produce steam, it takes a certain time for the heat to reach the water. This time lag is unimportant in the steady state (i.e., when there are no load changes) since it makes no difference if the steam consumed now was vaporized by fuel burned, say, a minute ago. The picture is totally different with a variable demand, however, since then there may be insufficient steam when needed or too much when not (i.e., excessive steam pressure).

The point is that, if the fuel control valve is overreacting, it may limit pressure swings, but the fuel/air ratio is being constantly upset, and this lowers the efficiency of combustion.[1]

The ideal situation is, of course, to maintain a steady state, but this is possible only part of the time, since steam demand is controlled by the process, not by the boiler. What then is an "acceptable" level of dynamic change? A closer look at the furnace dynamics may suggest criteria.

If a step change is made, manually, in the fuel control valve position, while the steam demand stays constant, the behavior in Fig. 4.1 will be observed. The output steam pressure response is that of a second-order linear differential equation in which T_f is one of the time constants. T_f, the furnace time constant, is usually between 10 and 20 s for liquid and gas fuels, higher for solid fuels.[2]

Why is the furnace time constant so important? It tells at what rate the heat is being absorbed. From a control point of view, theory says that for a first-order linear system, gain decreases with an increasing frequency in the input. If the period equals the time constant, an energy loss of 3 dB or 0.7 of the output occurs for a unitary input change.

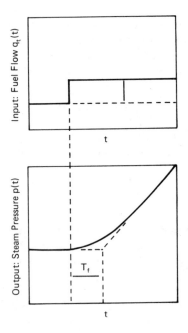

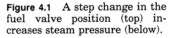

Figure 4.1 A step change in the fuel valve position (top) increases steam pressure (below).

To check for an overreacting fuel control valve in an existing installation, observe carefully the valve position and the fuel flow meter. If they are oscillating or constantly changing, it is an indication that the system is overreacting and wasting fuel by constantly upsetting the fuel/air ratio. Assure yourself that the flow meter is oscillating *because* of the control valve movement by going over to manual control, thereby locking the valve position.

Overdamping the valve's response will show excessive steam pressure swings.

Evaporator time constant

The boiler main time constant, or evaporator section constant,[1] T_e characterizes the steam accumulated in the water contained by the drum and part of the tubes. This constant is a design parameter and describes the ability of the boiler to withstand load changes. Typical values range from 200 to 400 s for drum boilers, less for once-through types.[1,2,3]

A low T_e value means faster steam pressure response to load changes (steps and ramps), while a high time constant has better regulating behavior for average pulsating or oscillating demand.

To determine the evaporator time constant, a simple test can be performed (Fig. 4.2). While keeping the fuel flow constant (controller in manual) increase or decrease steam flow by, say, 10 percent; then, start reading the pressure from an accurate pressure gauge, every 10 s for about 1 min. Use the following equation to determine the time constant:

$$T_e = \frac{\Delta q \, p_0}{q_{\max}\Delta p/\Delta t}$$

where Δq = change in steam flow over Δt
p_0 = steam pressure set point
$q_{\max}$ = boiler steam capacity
Δp = change in pressure over Δt
Δt = time increment (10 s)

Take the average of the calculations for the 10-s increments. [T_e can also be determined from the furnace response curve (Fig. 4.1).] Find the change in pressure Δp for a given time interval Δt. Use those values in the T_e equation, along with the corresponding change in fuel flow Δq_f, which replaces change in steam flow Δq in the equation. An accuracy within ± 10 percent in the determination of T_e is acceptable.

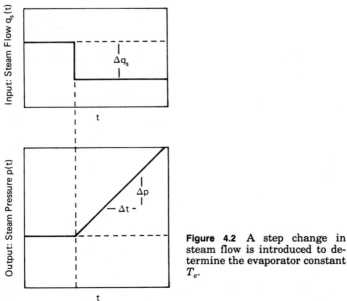

Figure 4.2 A step change in steam flow is introduced to determine the evaporator constant T_e.

Optimum boiler load response

From the curves in Fig. 4.3, we see how a large T_e limits the excursion of $\Delta p(t)$, the steam pressure. See also the mathematical model derivation in the next section.

In the plots of $\Delta p(t)$ and fuel flow $g_f(t)$, two things are especially important:

1. How fast does $q_f(t)$ change (maximum slope)? The highest slope for $q_f(t)$ is at the origin, and is 45°. That is, it would take $\Delta q_f(t)$ ω seconds at the initial rate to reach its final value, or 3 times the furnace time constant (30 to 45 s). This is pretty slow and therefore convenient.

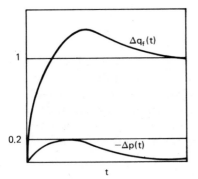

Figure 4.3 Binomial step response for third-order boiler load under PI control. (See the mathematical model derivation.)

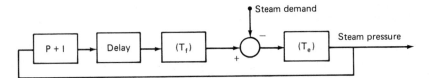

Figure 4.4 Control loop with air-before-fuel delay.

2. What is the maximum deviation for the steam pressure $\Delta p(t)$? To answer this we ask another question: How does the sluggish response of $q_f(t)$ affect the steam pressure? The maximum deviation will reach 0.84 of the disturbance over $T_e \omega$ seconds. For a 10 percent load change (which is typical), $T_e = 200$ (which is low), and $T_f = 15$, the deviation Δp_{max} is less than 2 percent, which is very good.

Boiler load control loop settings

For a PI-type controller and a natural circulation water-tube boiler, set the gain at approximately 10 ($G = T_e/3T_f$) and integral time at 100 s ($T_i = 9T_f$). Keep in mind that these are per-unit values and that they have to be corrected according to the calibration range of the transmitters, controller, and control valve.

Fuel/air ratio control

Some boilers have combustion-air controllers that operate independently of the actual load control loop, so that airflow can respond immediately to any change in fuel flow.

Other systems require an excess of air before they increase fuel flow. This will of course introduce an additional delay in the response, which can be simulated for modeling purposes by a first-order delay that is active only during load increases (Fig. 4.5).

The order of the system is now increased by 1 and the dynamics are slower. A similar model analysis can be done to determine the actual response, using the third-order controller settings suggested before. The air-controller delay time can be determined approximately, by observing it during operation.

Parallel boiler operation

If more than one boiler is supplying steam to the same header, every unit should participate equally in responding to load demand to reduce individual swings. However, it is possible that instabilities may occur, such as poor load distribution or one boiler taking over all the load.

For example, if the boilers have their own integration-time controller, each one will try to correct pressure on its own, with the fastest one succeeding and stealing the load from the others. To avoid this conflict, a unique controller can be used, with one output signal commanding the fuel valve of each boiler. This approach lumps the load control as if it were only one.

Boilers should be functionally the same in this case. Flexibility can be gained if each boiler has its own proportional controller (gain only) but one integral controller overall (Fig. 4.5)

Feedforward load control

Since feedforward will anticipate any steam pressure swing, it may seem attractive from a control point of view. However, as can be shown from model analysis (see below), this means bypassing the steam time constant (and its virtue of absorbing pressure fluctuations). It may prove necessary, though, when operating parallel boilers or using solid fuel, to introduce a feedforward signal to ensure proper load distribution. It should be avoided in single-boiler operations that burn oil or gas.

If a decision is made to use feedforward, note that if the feedforward signal is not properly delayed, each change in steam load will generate an immediate change in fuel flow, and a decrease in efficiency therefore should be expected.

With feedforward, a simple solution with binomial forms is not possible. Analysis of a dynamic feedforward model (Fig. 4.6) suggests that the incoming signal should bypass the controller and act on the fuel valve directly with a time delay in the order of $T_i[K/(1 + sT_i)]$ and

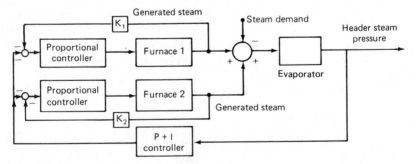

Figure 4.5 Parallel boiler. Generated steam by a boiler in parallel can now be measured and the signal used to improve load distribution.

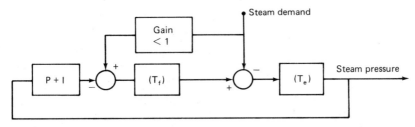

Figure 4.6 Feedforward control.

gain K less than 1. (If there is more than one boiler, K will be the percentage load that the boiler will pick up.)

Residual oxygen control

The main purpose of this section is to show ways of making residual oxygen air-trim control more feasible by slowing down the dynamics of the system. There is not much sense in trying to control residual oxygen during changing load conditions.

Steam accumulator fire-tube boiler option

In order to keep steam generation constant with a variable demand, some excess steam can be stored in a steam accumulator[4] for use when demand increases over generation. This may not always be economically practical or feasible. Fortunately, steam generators (i.e., boilers) can also store steam, and this should be kept in mind during the design stage so that a higher heat-storage capacity (i.e., higher T_e) can be allowed if necessary.

Given a permissible pressure drop of, say, 10 percent, the following values are typical for boiler storage[4]:

- Fire-tube boilers at up to 150 lb/in^2 operating pressure have sufficient storage capacity to increase the boiler output by 500 percent for 1 min or about 8 percent for 1 h ($T_e \approx 3000$ s).

- Water-tube boilers at 450 lb/in^2 operating pressure can increase output only about 35 percent for 1 min or 0.6 percent for 1 h ($T_e \approx 200$ s).

Mathematical model of a boiler load control loop

If fuel flow is kept constant, steam pressure will start changing at a constant rate for a step change in demand:

$$\left[\frac{\Delta q(t)}{q_{max}}\right]_{steam} = -T_e\frac{\dot{p}(t)}{p_0}$$

where q represents flow, p_0 = set-point steam pressure, and $\dot{p}(t) = dp(t)/dt$, the first time derivative of steam pressure. The total steam flow is the net total of demand steam flow (subscript ds) and generated steam flow (subscript gs):

$$\Delta q(t)_{steam} = q(t)_{gs} - q(t)_{ds}$$

A step change in only the fuel flow will cause an exponential change in the steam flow:

$$\left[\frac{\Delta q(t)}{q_{max}}\right]_{fuel} = \left[\frac{q(t)}{q_{max}}\right]_{gs} + T_f\dot{q}(t)_{gs}$$

where $\dot{q}\,(t)$ is the first time derivative of flow (in this case, generated steam flow). So the steam pressure will behave as follows:

$$p_0\left[\frac{\Delta q(t)}{q_{max}}\right]_{fuel} = T_e\dot{p}(t) + T_fT_e\ddot{p}(t)$$

The transfer function is then

$$Q_f(s) = sT_eP(s) + s^2T_fT_eP(s)$$

The block diagram for the model is in Fig. 4.7.

Single-boiler PI control

The input to a PI controller for a single-boiler system is steam demand and outputs are fuel flow and steam pressure. The fuel flow should try to follow the steam demand (a servo problem) with the limitations that were mentioned at the beginning of the chapter. Steam pressure has to stay fairly constant (a regulator problem) when changes take place.

Now consider a fast-responding control valve actuator. This simplifies the model solution by eliminating a time constant (i.e., if the valve time constant is low compared to the other time constants, say, 3 s). Then the ideal controller transfer function is

$$PI = \frac{1 + sT_i}{sT_i}G$$

The overall transfer functions are

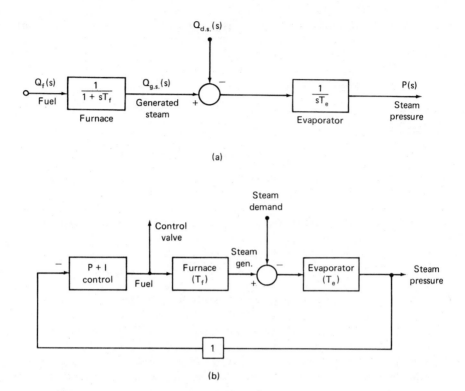

Figure 4.7 (a) Boiler transfer function; (b) load-control loop.

$$\frac{Q_f(s)}{Q_s(s)} = \frac{(G/T_e)(s + 1/T_i)(s + 1/T_f)}{s^3 + s^2/T_f + sG/T_fT_e + G/T_fT_eT_i}$$

$$\frac{P(s)}{Q_s(s)} = \frac{\dfrac{s}{T_e}(s + 1/T_f)}{s^3 + s^2/T_f + sG/T_fT_e + G/T_fT_eT_i}$$

If we equate the characteristic equation to a standard form of the type:

$$s^3 + a\omega s^2 + b\omega^2 s + \omega^3$$

where ω = natural frequency and a, b = constants, we get the following controller setting expressions:

$$\omega = \frac{1}{aT_f} \qquad T_i = abT_f \qquad G = \frac{bT_e}{a^2T_f}$$

Observe that T_f alone determines the natural frequency of the boiler in this arrangement.

Load control transient response

For the binomial standard form $(s + \omega)^3$, which yields critically damped responses,

$$a = b = 3$$

$$\omega = \tfrac{1}{3} T_h = \frac{G}{T_e} \qquad T_i = 9T_f \qquad G = \frac{T_e}{3T_f}$$

For fuel flow, the transient response to a step load change Q_s/s is

$$Q_f(s) = \frac{\omega Q_s(s + 3\omega)(s + \omega/3)}{s(s + \omega)^3}$$

and the inverse Laplace transform (time domain) is

$$\Delta q_f(t) = \Delta q_s\{1 + [\tfrac{2}{3}(\omega t)^2 - 1]e^{-\omega t}\}$$

where Δq_s is a step change in steam flow demand.

For steam pressure, the transient response to the step load change is

$$P(s) = -\frac{Q_s(s + 3\omega)}{T_e(s + \omega)^2}$$

and the inverse Laplace transform is

$$\Delta p(t) = -\frac{\Delta q_s}{\omega T_e} \omega t_e^{-\omega t}$$

Energy-Efficient Dynamic Control

How efficient is linear feedback control? Does it require any energy to run an *ideal* control loop? The answer is "yes," a perfect feedback control loop dissipates energy, regardless of what the application is. This becomes obvious from an electrical resistance-capacitance (RC) circuit representation of a feedback loop (Fig. 4.8).

A feedback control loop must dissipate energy in order to remain stable, otherwise it would oscillate. The energy dissipated in response to a step input V_i can be calculated by integrating the transient response energy function from 0 to ∞:

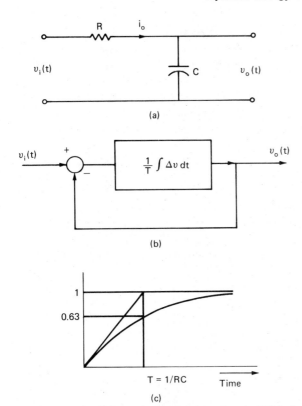

Figure 4.8 An RC circuit showing (a) network, (b) block diagram representation, and (c) step input response.

$$W = \int_0^x i(t)^2 R \, dt = \int_0^x \left[\frac{V_i(0)}{R} e^{-t/T} \right]^2 R \, dt$$

(The terms are defined in Fig. 4.8.) The integration yields

$$W_{in} = \frac{P_{in(t = 0)} T}{2}$$

where $P_{in(t = 0)} = V_i(0)/R^2$ is the instantaneous inrush load. In other words, it requires one-half the inrush load times the loop time constant to bring the system to a new state. The other half is required to bring the system back to its original state!

Observe that it is irrelevant whether the energy flows in or out. The closed-loop system will consume—i.e., dissipate—energy in order to change state! *Additional energy is dissipated in a feedback loop each time there is a change of state, regardless of flow sign.*

Observe also how the energy consumption depends on the time constant. Feedback loops store energy in direct proportion to their time constants. Therefore, feedback loops with large time constants absorb load fluctuations better but are more inefficient to changes in the input, and vice versa. Actually, there is no mystery here, since we already know that an alternating current dissipates energy, regardless of the fact that the average voltage and current are zero. So we could have found the energy dissipated in a closed loop by a root-mean-square (rms) calculation.

Set-Point-Change and Load-Change Energy Efficiency Analysis

Let's analyze the configuration in Fig. 4.9, which presents an electrical circuit analogy for a load control application. The purpose is to maintain output voltage $v_o(t)$ as nearly constant as possible, despite

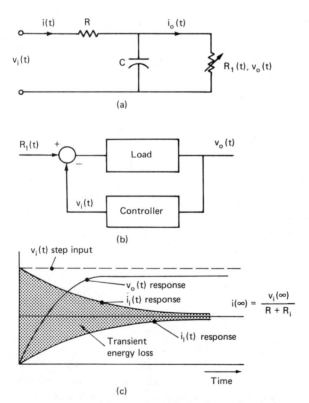

Figure 4.9 An RC circuit with load on the output. (a) Network with load; (b) block diagram representation; (c) load step response.

the random fluctuations of load resistance $R_i(t)$ and its associated current $i_o(t)$. Resistance R is constant and input voltage $v_i(t)$ can be manipulated by a controller to maintain $v_o(t)$ constant.

This is an electric equivalent model of a heat-exchanger control loop: $v_i(t)$ represents the heat source, R the exchanger well resistance, and $R_i(t)$ the fluctuating load applied to the heat exchanger.

What happens to the energy when:

1. There is a step change in $v_i(t)$ (control input)?
2. There is a step change in $R_i(t)$ (load input)

In case 1, the response has two components: a step response and exponential response in the voltage across capacitance C. The transient energy loss is the energy integral of the shaded area in Fig. 4.9b.

$$W_{\text{loss}} = \int_0^x i_c^2(t)\, R\, dt$$

where

$$i_c = C\frac{dv_o(t)}{dt} = C\dot{v}(t)$$

$$v_0(t) = i_l(t)\, R_l$$

$$i(t) = i_c(t) + i_l(t) = C\dot{v}_o(t) + \frac{v_o(t)}{R_l}$$

$$v_i(t) = R_i(t) + v_o(t)$$

$$v_i(t) = R\left[C\dot{v}_o(t) + \frac{v_o(t)}{R_l(t)}\right] + v_o(t)$$

$$v_i(t) = RC\dot{v}_o(t) + \left[\frac{R}{R_l(t)} + 1\right]v_o(t)$$

$$RC = T$$

$$V_i(s) = \left(sT + \frac{R + R_l}{R_l}\right)V_o(s) \qquad \text{where } R_l(t) = R_l$$

$$\frac{V_o(s)}{V_i(s)} = \frac{1}{sT + A} = \frac{1}{A}\left(\frac{1}{sT/A + 1}\right)$$

$$V_o(s) = \frac{v_i(0)}{As} \left(\frac{1}{sT/A +1}\right) \frac{V_i(0)}{A} (1 - e^{-t/(T/A)}) \qquad \text{where } v_i(0) = \text{step input}$$

$$v_o(t) = \frac{v_i(0)}{A} (1 - e^{-t/(T/A)})$$

$$\dot{v}_o(t) = \frac{v_i e^{-t/(T/A)}}{T} \qquad V_i(0)$$

$$T' = \frac{T}{A} = \frac{R'}{R}$$

$$\int_0^x RC^2 (\dot{v}_o)^2 \, dt = \int_0^x RC^2 \frac{v_i^2(0)}{T^2} e^{-2t/T'} dt$$

$$= \frac{v_i^2(0)}{R} \int_0^x e^{-2t/T'} dt = \frac{v_i^2(0)}{R} \frac{T'}{(2)}$$

$$= I^2(0)R\frac{T'}{2}$$

$$= W_{\text{loss}}$$

The choice of a value for R thus strongly affects transient and steady-state energy loss W_{loss}. Consequently R should be much smaller than R_l, and this will also reduce the time constant T.

The correct interpretation of the W_{loss} equation is that, once energy is stored in the system, it should be kept at the same level by slow control input, i.e., $v_i(t)$ changes. That is why slow systems—i.e., those with large time constants—are less efficient than faster ones when they are subjected to changes in set point. On the other hand, some systems *must* follow continuous set-point changes, and in those cases time constants should therefore be kept small to save energy. In case 2 a step change of $R_l(t)$ load will only cause a gradual change of $v_o(t)$, which implies a gradual change instead of step change for $v_i(t)$ by the controller, and the transient energy loss will be substantially smaller.

References

1. G. Quazza, "Role of Power Station Control in Overall System Operation," *Brown Boveri 1971 Symposium Proceedings*, E. Handschin (ed.), Elsevier, New York, 1972.
2. G. Klefenz, "Die Regelung von Dampfkraftwerken," Hocshultaschenbucher 549/549a, Bibliographisches Institüt Manheim.
3. M. Polonyi, "Boiler Dynamic Load Control and Optimization," *Control Engineering*, September 1985.
4. Walter Goldstern, *Steam Storage Installations*, 2d ed., Pergamon Press, New York.
5. J. D'Azzo, *Feedback Control Systems*, chap. 17, McGraw-Hill, New York.

Chapter 5

Drum Level Control in High-Pressure Steam Boilers

Drum Level

The emphasis in steam drum level control is not on accuracy or efficiency but rather on limiting the level swings within a safe margin during load changes, no matter how large these may be, so that dangerous levels will never be reached. Excessively high levels are particularly dangerous for steam turbines because water can be drawn into the turbine and damage the turbine blades (or "buckets"). Excessively low levels expose water-cooled sections of the boiler; overheating can then destroy them.

High-pressure steam boiler level control is a rather complicated process. One reason is that the initial level reaction in a steam drum goes against expectations; that is, level *increases initially when load increases and vice versa*. This is caused by the sudden drop in steam pressure in the drum, which releases extra bubbles from the water, causing it to "swell" (Fig. 5.1). The opposite effect "shrink," takes place when load decreases, since the steam pressure then suddenly increases (Fig. 5.2). Changes in feedwater flow cause similar disturbances; an inrush of relatively cool water produces shrink, and a sudden decrease in flow produces swell. These effects limit the allowable response speed of the feedwater/level control valve. The level set point is chosen so that shrink (or swell) is minimal (Figs. 5.3 and 5.4). As conditions settle down, the logical trend starts taking over, and level starts moving in the "right" direction—that is, level drops as load continues to increase and vice versa. Figure 5.5 shows a block diagram of the overall boiler control system and a transfer function model of the level controller.

The swelling and shrinking effects in steam drums can be modeled

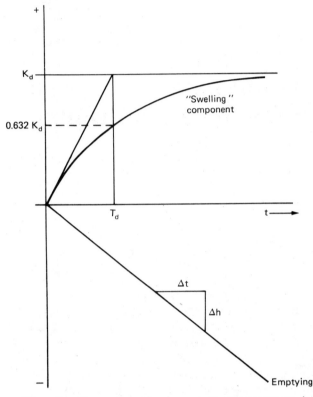

Figure 5.1 Steam drum level response components to increased steam load or decreased water flow.

as a first-order time constant[1] with a negative trend superimposed on the level integration constant. This is called a *non-minimum-phase* system, i.e., one which affects unfavorably the stability of the system.

In order to limit these excursions of level, three-element level control has been designed (Fig. 5.6). In control parlance, this means that a feedforward signal has been added to the control loop.

The three elements are the steam flow transmitter, the level transmitter, and the feedwater flow transmitter. The feedforward signal comes from the steam flow transmitter, i.e., it reflects steam boiler load. The third element, the feedwater flow transmitter, adds to the accuracy of the feedforward signal by smoothing out any feedwater flow control element nonlinearities under different flow and pressure conditions.

So, basically, we have two control loops superimposed on only a sin-

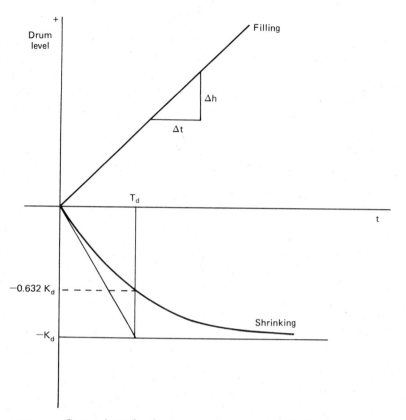

Figure 5.2 Steam drum level response components to decrease in steam load or increase in feedwater flow.

gle control element, the level control element (a valve or variable speed pump), and this creates a basic conflict. Which loop should be given priority? Level, of course, should override the steam flow signal if there is an opposing trend between these two. However, when the steam drum level is close to its set point, it is *change of flow* that has an immediate effect on level and becomes the controlling signal.

On the other hand, imagine a situation in which the level is high and the load is high. The level controller wants to close the valve while the flow controller wants to keep it open. We still want the level to come down, so the level controller has to override the steam flow signal. This eventually will occur as a consequence of the reset action of the level controller.

The factors that intervene in steam drum level control are complicated and often contradictory. Settings therefore have to be determined carefully.

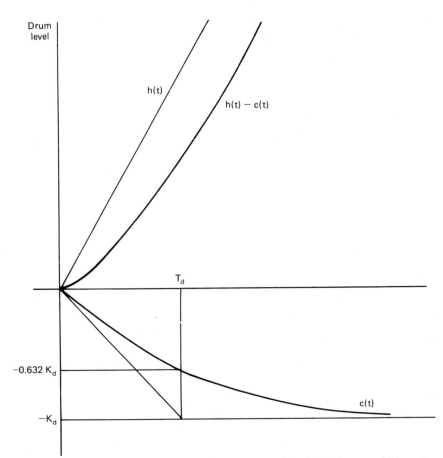

Figure 5.3   Steam drum level change without observable shrink because $h(t) > c(t)$ over the entire range. Drum level = $h(t) - c(t)$.

Both loops act on the control valve independently, with the flow controller active when the level is close to its set point, and the level controller overriding the flow controller when the level is too high or too low (Fig. 5.7). To implement this arrangement, the level controller must have slow response—a binomial response or slower—to allow the level to drift somewhat so as not to create thermal shocks by abrupt opening and closing of the feedwater flow control element. The flow controller should have tight settings in order to respond immediately to changes in steam load, thereby offsetting the swell and shrink phenomena.

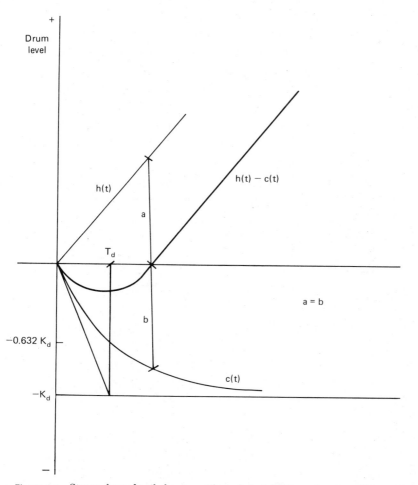

Figure 5.4   Steam drum level change with real shrink. Drum level $= h(t) - c(t)$.

The level controller gain setting ought to be limited to provide the overdamped response required in a tank-filling operation (see the next section in this chapter).

Steam drum level controllers that cannot keep the level constant are not designed or properly adjusted. Adjustment of controllers sometimes requires repeated tests, which are not always possible or easy to run on large boilers. However, if the control loop is implemented in a DCS, it is a fairly simple matter to redesign it, or change any setting, and a better response is possible because of the flexibility that a DCS offers. If the control loop is implemented with pneumatic instrumen-

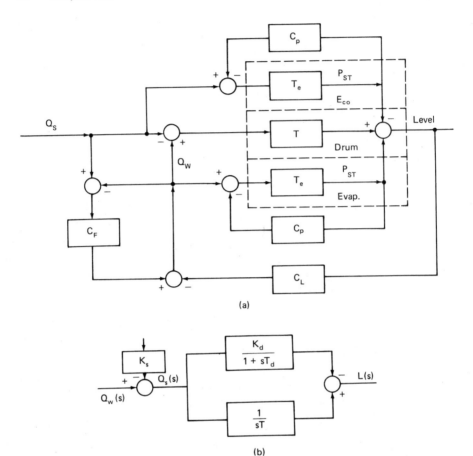

Figure 5.5 (*a*) Steam drum level control model block diagram. (*b*) Steam drum level simplified model block diagram. C_F: feedwater flow controller, C_L: level controller, C_P: steam pressure controller, T: drum integration constant, T_e: evaporator integration constant, Q_S: steam load, Q_W: feedwater flow, P_{ST}: steam pressure, E_{CO}: economizer section, Evap: evaporator section, K_D: "shrink-swell" effect gain.

tation, calibration is not as accurate. Pneumatic ranges are more cumbersome to set and change, and response is much slower.

Steam Drum Level Set Point

Swelling and shrinking can be minimized if the level set point is correctly determined. This is strictly a boiler design parameter. In a correctly chosen and adjusted level control loop, swelling and shrinking will be "transparent," they will translate into no observable changes of level at all.

Boiler time constants change little from one to another within the

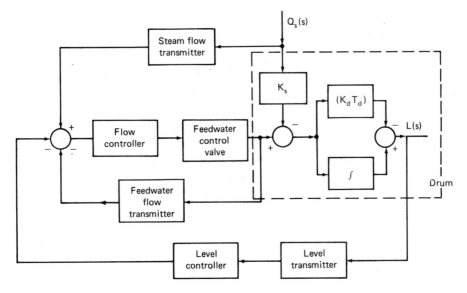

Figure 5.6   Three-element drum level control block diagram.

same model type, i.e., water-tube, waste-heat, once-through, and so forth. Once the closed-loop expression has been determined, it can be equated to a standard form of known response to obtain the controller settings that will do the job.

Making maximum feedwater flow a "base" value ensures that the level control valve has gain = 1, thereby simplifying its representation. It will still have a time constant associated with the valve actuator and positioner.

The transmitter gains will be

$$\text{Level transmitter gain} = \frac{\text{maximum inferior normal level change}}{\text{level transmitter span}}$$

$$\text{Flow transmitter gain} = \frac{\text{maximum feedwater flow}}{\text{flow transmitter span}}$$

In swelling and shrinking, a 10-s time constant is used in a first-order model representation. The level transmitter signal will be pressure- and temperature-compensated so that the associated changes of density will not affect the true level signal.

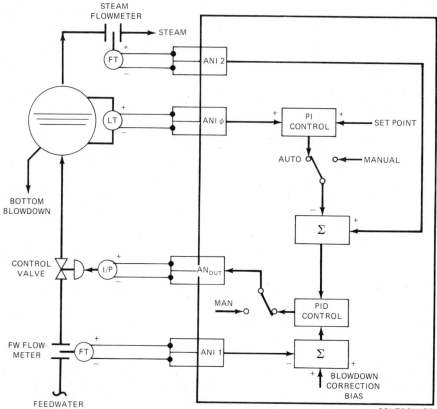

Figure 5.7 Three-element boiler drum level control schematic. *Note:* Steam and feedwater flow meters must have the same range, but feedwater flow must account for bottom blowdown and other purges, through a manual bias signal or otherwise.

Mathematical Model Analysis and Synthesis

Flow control loop

The feedwater flow control loop model is analyzed as follows, with reference to Fig. 5.8. The flow control transfer function is

$$\frac{Q_W(s)}{Q_S(s) - L(s)} = \frac{1 + \text{PID} \dfrac{1}{(1 + sT_v)}}{1 + \text{PID} \dfrac{1}{(1 + sT_v)} \dfrac{1}{(1 + sT_t)}}$$

where T_v = control valve time constant (about 5 s)
   T_t = feedwater transmitter time constant (< 1 s)

$$\text{PID} = G(1 + 1/sT_i + sT_d) \qquad \text{where } T_d = 0$$

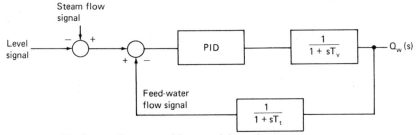

Figure 5.8 Feedwater flow control loop model synthesis.

The characteristic equation is

$$1 + G\left(1 + \frac{1}{sT_i}\right)\frac{1}{(1 + sT_v)}\frac{1}{(1 + sT_t)} = 0$$

which, after algebraic manipulation, becomes

$$s^3 T_i T_v T_t + s^2 T_i (T_v + T_t) + sT_i(1 + G) + G = 0$$

or

$$s^3 + s^2\left(\frac{1}{T_v} + \frac{1}{T_t}\right) + s\frac{1 + G}{(T_v T_t)} + \frac{G}{T_v T_t T_i} = 0$$

Then the terms in the standard-form equation and corresponding controller setting equations are

$$a\omega = \frac{1}{T_v} + \frac{1}{T_t} \qquad \omega = \frac{1}{a}\left(\frac{1}{T_v} + \frac{1}{T_t}\right)$$

$$b\omega^2 = \frac{1 + G}{T_v T_t} \qquad G = b\omega^2 T_v T_t - 1$$

$$\omega^3 = \frac{G}{T_v T_t T_i} \qquad T_i = \frac{G}{\omega^3 T_v T_t}$$

For $T_v = 5$ s and $T_t = 1$ s, the feedwater controller settings are as shown in Table 5.1.

TABLE 5.1 Feedwater Controller Settings—Third-Order Model

Criterion	a	b	ω, rad/s	G	T_i, s
Critically damped (slow)	3	3	0.4	2.2	6.9
Butterworth (intermediate)	2	2	0.6	2.6	2.4
ITAE zero displacement (fast)	1.75	2.15	0.69	4.1	2.5

Notes: These settings can be used for any flow control loop.
Steam and water flow meters have the same range.
A slight error may occur because of unaccounted flow for bottom blowdown and other purges.

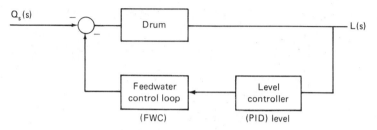

Figure 5.9 Level control loop.

Level control loop

The level control loop is analyzed as follows, with reference to Fig. 5.9. The level control transfer function is

$$\frac{L(s)}{Q_S(s)} = -\frac{1 + \text{drum}}{1 + \text{drum} \times \text{FWC} \times \text{PID}_{\text{level}}}$$

The transfer functions for the drum and feedwater control loop are

$$\text{Drum} = \frac{1}{sT(1 + sT_d)} \qquad \text{for } K_d T = T_d$$

$$\text{FWC} = 1$$

Typical time constants are $T = 20$ s and $T_d = 10$ s.
   The characteristic equation is

$$1 + G\left(\frac{1 + sT_i}{sT_i}\right)\frac{1}{(1 + sT_d)(sT)} = 0$$

Manipulation gives

$$s^3\, T_i T_d T + s^2\, T_i T + s\, GT_i + G = 0$$

or

$$S^3 + \frac{S^2}{T_d} + S\frac{1 + G}{(T\,T_d)} + \frac{G}{T_i T_d T} = 0$$

The standard-form equation is

$$s^3 + a\omega\, s^2 + b\omega^2\, s + \omega^3 = 0$$

The terms in the standard-form equation and corresponding controller setting equations are

$$a\omega = \frac{1}{T_d} \qquad \omega = \frac{1}{aT_d}$$

The steam drum

$$b\omega^2 = \frac{G}{T\,T_d} \qquad G = b\omega^2 T T_d$$

$$\omega^3 = \frac{G}{T\,T_d T_i} \qquad T_i = \frac{G}{\omega^3 T T_d}$$

Representative values of level controller settings for fastest response are as shown in Table 5.2.

TABLE 5.2   Steam Drum Level Fastest Controller Settings*—Third-Order Model

Criterion	a	b	ω, rad/s	G	T_i, s
Critically damped	3	3	0.033	0.66	92

Note: G is the *loop gain* and *must* be corrected for level transmitter gain.
*Slower-response settings can be obtained by increasing a and b arbitrarily.

Filling Scale Tanks Using Modulating Control Valves

When scale tank filling (Fig. 5.10) must be done accurately, analog control techniques can be used. An accuracy of 1 part in 1000 will result in more efficient use of raw materials and better product quality.

Accuracy with a controlled on-off valve may fall short of this target. Take, for example, a scale tank that has to be filled to 600 gal by a 150-gal/min pump. Throttling to an accuracy of 1 part in 1000 means accuracy within

$$\frac{600}{1000} = 0.6 \text{ gal}$$

At a 150-gal/min filling rate this accuracy equates with a maximum time delay of

$$\frac{0.6 \text{ gal}}{150 \text{ gal/min}} \times 60 \text{ s/min} = 0.24 \text{ s}$$

to move the valve from fully open to fully closed. This time lapse might not seem large in comparison with actuator or controller delay times. But if we start adding the time lags for each element in the control chain, the total delay becomes large. We can take dead time into account and compensate for it, but just a small *variation* in total dead time will exceed our required accuracy.

Another disadvantage of the on-off valve method is that it requires a dedicated programmable controller because of the high switching

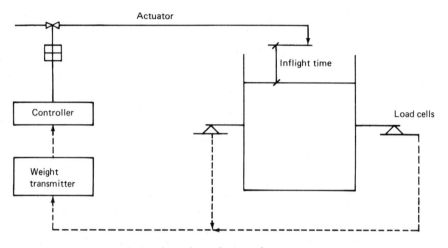

Figure 5.10   Level control of scale tanks and steam drums.

velocities involved. This may be a problem in distributed process control systems (DCS) because the interfacing between the process controller and the programmable controller degrades some of the original capabilities of either system. Two cascaded on-off valves can also be used, but let's explore the use of a single modulating control valve.

Let's assume the valve has a turndown ratio of at least 1:50, which is acceptable by today's standards. Then

$$\frac{150 \text{ gal/min}}{50} = 3 \text{ gal/min}$$

is the minimum controllable flow rate. The time lapse to fully close the valve is then

$$\frac{0.6 \text{ gal}}{3 \text{ gal/min}} \times 60 \text{ s/min} = 12 \text{ s}$$

This is already 50 times better than on-off control.

Now, let's be pessimistic and assume the following time lags:

Control valve	3 s
Weight transmitter	1 s
Inflight	3 s
Total	7 s

This gives us about + 50 percent allowed inaccuracy in valve-closing time, which is significantly better than on-off control also. Another advantage is that we can use now the DCS controller with *lower* scanning rates and, therefore, eliminate the interface and special time-compensating programming required with the dedicated programmable controller arrangement.

Control Loop Parameter Tuning

The most simple approach is to use a pure gain controller (Fig. 5.11), for which the tank-filling time integration constant is

$$T_t = \frac{600 \text{ gal}}{150 \text{ gal/min}} \times 60 \text{ s/min} = 240 \text{ s}$$

The transfer functions are

$$\frac{H(s)}{I(s)} = \frac{K}{sT_t + K}$$

and

$$\frac{Q(s)}{I(s)} = \frac{sKT_t}{ST_t + K}$$

where $I(s)$ = step input fill command, K = controller gain, and T_t = tank integration constant. The time responses are

$$h(t) = 1 - e^{-tK/T_t}$$

$$q(t) = e^{-tK/T_t}$$

However, a controller gain higher than 1 means that the tank is filling at a higher rate than 150 gal/min, which is impossible. Actually, a controller gain higher than 1 means that the controller will not start closing the control valve from the very beginning, and this results in a reduction of the calculated time constant. The corrected integration constant is

$$T = \frac{T_t}{K}$$

We also can see that the shortest filling time will be given by the highest gain, so why not make the gain infinity. If we do this we will

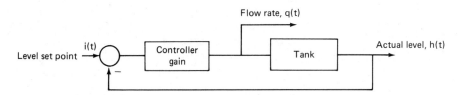

Figure 5.11 Proportional band controller block diagram.

have gone back to on-off valve control, therefore there is a limit to the amount we can increase K.

We can calculate this limit if we take into account all the time lags in the system. For example:

Controller	1-s dead time
Inflight	2-s dead time
Control valve	3-s time constant (first-order)
Weight transmitter	1-s time constant (first-order)
Settling time	1-s time constant (first-order)

We could ignore the dead time, since we can compensate for it by using a Smith predictor in our control algorithm, but for simplicity let us assume only one first-order time constant of, say, 10 s for all the control elements. (We can simulate this composite time constant as part of our control algorithm.) The transfer functions for the composite control loop (Fig. 5.12) are

$$\frac{H(s)}{I(s)} = \frac{K}{sT(1 + sT_c) + K} = \frac{K/TT_c}{s^2 + \dfrac{S}{T_c} + \dfrac{K}{TT_c}}$$

and

$$\frac{Q(s)}{I(s)} = \frac{SKT}{sT(1 + ST_c) + K} = \frac{SK/T_c}{s^2 + \dfrac{S}{T_c} + \dfrac{K}{TT_c}}$$

Since this is a tank-filling operation, we are not allowed any overshoot more than the overfill error. If our response has to be damped, the real roots in the denominator of the transfer functions will give the information we need. This means $K < 2.5$. Let's take $K = 2$ and see what our filling time is. $K = 2$ also means that, during the first half of the filling process, our valve will be fully open; therefore, during the first half:

$$t_i = \frac{300 \text{ gal}}{150 \text{ gal/min}} = 2 \text{ min filling time (first half)}$$

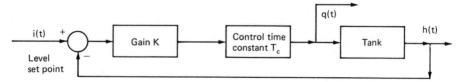

Figure 5.12   Composite control loop for tank filling.

Finding the second-half filling time is more complicated. The transfer function is

$$\frac{H(s)}{I(s)} = \frac{2}{s(120)[1 + s(10)] + 2} = \frac{\frac{1}{600}}{s^2 + \frac{s}{10} + \frac{1}{600}}$$

The time response is

$$h(t) = 1 - 1.36e^{-0.021t} + 0.36e^{-0.08t}$$

For a total error of 0.001, this means a 0.002 error in the control range chosen, and we get

$$t_2 \approx \frac{-1}{0.021} \ln \frac{0.002}{1.36} = 310 \text{ s} \approx 5 \text{ min}$$

This 5 min plus the 2 for full-flow filling makes a total of 7 min, which is almost twice the time for filling the entire tank at full flow. Although still not a significant time in the overall process, this is the price we pay for our increased accuracy.

We also see that gain adjustment should be enough to take us to our goal. The final gain setting should be field-adjusted to attain the highest filling rates possible without overshooting our error band.

It is also possible to attain a lower filling time by adding reset and rate action in our analysis. In this case, accurate modeling is required to prevent overshoot. Other control schemes are also possible.

A significant feature of process-distributed control systems is that sophisticated control algorithms are available as inexpensive software. Therefore schemes with intensive use of algorithms should take precedence over hardware options.

Finally, the only limitation in accuracy is not a valve problem but a sensor one. Find the overall accuracy by adding the relative error of each sensor.

1. G. Klefenz, "Die Regelung von Dampfkraftwerken," Hocshultaschenbucher 549/549a, Bibliographisches Institüt Manheim.

Chapter 6

Cogeneration Power Plant Dynamics and Control

Introduction

From a technical point of view, a cogeneration plant is a small power plant, but to utilities cogeneration in actuality is "the competition." Electric utilities do not compete among themselves; they are monopolies within their own territories—except when a utility must connect to a cogeneration plant.

It is not easy to beat utilities at their own game, even when they are receptive to (that is, need) cogenerators to increase their capacity. In order to successfully compete with power delivered by utilities a series of conditions must be met.

The following are some of the requirements to make a "cogen" plant economically viable:

- The utility is receptive to applications for cogeneration; that is, it does not insist on harsh clauses and penalties.

- The cogeneration plant may set its own loading schedule (a base load for oil or gas, an erratic load for all other fuels).

- Waste heat from engines and turbines can be utilized to supplant other fuels.

- The utility will provide 100 percent on-line backup and standby power.

- Emergency or stand-alone generation is not a requirement.

Financing for a cogeneration plant must contemplate the following:

- Capital equipment investment
- Engineering and construction

- Operation (personnel and fuel)
- Maintenance (service and parts)

It is common for planners to get carried away by claims about cogeneration's benefits, but the facts remain that the road to cogeneration is full of pitfalls and cogeneration is hard to justify most of the time. Remember: it can always be made to look good on paper.

Utility Contracts

For a cogeneration plant to be economically justifiable, it must be able to operate on-line with the utility grid. The utility will then provide:

- Start-up capability
- On-line backup
- Energy buy-back

Sometimes, however, the economic factor is secondary, and a cogen plant *is required*, e.g., for standby emergency power in hospitals or for elevator service in skyscrapers during blackouts. In these cases the emergency plant can be used on a daily basis to lower utility bills via cogeneration, and only the avoided energy cost has to be figured into the projected savings because engineering, construction, and operation and maintenance (O&M) are part of a different budget.

When utilities are required by law to accommodate cogeneration but are not receptive to it, then all kinds of excuses, obstacles, and contract clauses can be created to finally discourage the prospective cogenerator. Still, in the case of emergency power, it is important to be able to run the engines, preferably on a daily basis, under real load conditions, to become familiar with their operation when an unexpected blackout occurs. To get the utility's cooperation for these cases, at least, by allowing synchronization for a short period of time, it is very important to reduce the extra capital cost of backup and simplify load-transfer operations.

Utility contracts for cost of standby power and availability clauses can be especially detrimental to a project. In these cases, the utility is forcing the cogenerator to accept the fact that it cannot economically match the level of reliability that the utility service offers.

Types of Cogeneration Plants

Any known form of electric power generation, except for nuclear reactors, qualifies for cogeneration, provided their size remains small. For example:

- Diesel engines

- Gas turbines

- Steam turbines with boilers burning oil, gas (natural, topping, digester, etc.), or solid waste (garbage, wood, bagasse, car tires, cow chips, etc.)

- Combined cycles [gas turbine with heat-recovery steam generator (HRSG), and steam turbine combination]

- Small hydroelectric projects

- Wind "farms"

- Solar (e.g., photocells, heat)

- Geothermal

The most successful ones from an economic point of view are (1) those for which electricity is a by-product of heat load, as in the food and petrochemical process industries in general; and (2) those that can use process waste for fuel.

"Resource-recovery," or garbage-burning, plants are lucrative because they derive income twice, once to receive the garbage (tipping fee) and then to sell electric energy; i.e., not only is their fuel free, but *they get paid to burn fuel.*

Oil and natural gas, on the other hand, are very expensive to burn, and the project becomes viable only for utility-sized projects or if waste heat can be recovered and used in substantial amounts. An electric generation efficiency in the neighborhood of 30 percent leaves a 70 percent heat balance to be recovered for useful purposes.

Waste-heat recovery is cumbersome, and waste-heat boilers are not as efficient as those that burn fuel directly because, of course, the waste-heat temperatures are much lower. Besides, the amount of waste heat produced is dictated by the electric generator load, not by the heat demand; consequently *an additional source of heat must be provided to make up for thermal-load cycling.*

Load Cycling

Since electricity is the most valuable product to a cogenerator, the goal is to maximize its output. This makes cogeneration a base-load operation by default, and utilities are left with an increased burden of power-frequency, or "ripple," control. In other words, *cogeneration makes the network increasingly unstable unless it is tied to a regional load dispatch, and this reduces somewhat its economic viability.*

Load cycling increases thermal and economic losses substantially.

Waste-heat recovery many times makes a project economically viable. However, if the thermal load has its own cycle, an additional source of heat must be provided with very special features. A cycling backup heat source must be able to run continuously anywhere between 0 to 100 percent (0 to 100 percent rangeability) and operate efficiently throughout the range. Again, this is not an easy requirement to meet, since most fired boilers cannot sustain efficient furnace conditions at very low loads. Cycling heat sources need to have substantial heat storage capacity. For steam, the *fire-tube boiler* meets that requirement and *becomes an ideal buffer when steam pressures do not exceed 250 lb/in²*.[1]

Pressurized hot-water boilers, which actually do not boil, have enormous heat-storage capacity that makes them very efficient. The combination of pressurized-hot-water and low-pressure-steam heat exchangers makes an energy-efficient installation.

Cogeneration Plant Sizing

What is the optimum installed capacity for a power plant? Should it be based on maximum capacity, average capacity, or backup capacity? Short-term peak load or long-term base load?

For every kind of service, there are certain dynamics that will dictate the optimum size and characteristics, i.e., the "personality" of the generating plant, units, and service. These dynamics are based primarily on *cycling duty*, which translates into utilization factor, load factor, stability, transient response, efficiency, reliability, and backup. All of these characteristics must be considered and balanced somehow in order to obtain the optimum design and avoid unpleasant realities after the money has been spent. Sometimes the balance is straightforward, but usually hard choices must be made.

Island-Mode Diesel Plant Sizing Example

Take for example a diesel power plant whose first requirement is to provide electric power continuously 24 hours a day, 365 days a year because the utility cannot or will not interconnect with it (i.e., the plant must operate in the "island mode").

The plant has to meet the following requirements to operate successfully in the island mode:

- Satisfy maximum (peak) demand
- Function stably at minimum demand
- Maintain an on-line spinning reserve
- Provide hot and cold standby reserve

- Have adequate installed/spare capacity

Diesel engines have important advantages in this kind of service. They

- Have the most reliable record of any kind of power plant
- Can start cold and be on-line in less than a minute
- Can pick up 100 percent load instantaneously
- Can be overloaded for short·periods of time

On the other hand, diesel engines have disadvantages. They

- Burn expensive fuel
- Cannot be continuously loaded below 50 percent of capacity
- Require water cooling.
- Decrease efficiency together with reduction in loading.

Reliability considerations

In island mode operation, it is not economically feasible to have the same degree of reliability that a utility service has. So the first hard choice must be made: what kinds of outages can be tolerated and for how long.

Assume that a single diesel generator will be available 90 percent of the time for all possible scenarios. Then for two such engines, outage time would be 1 percent, that is 88 hours/year or 3.5 days/year. This is usually unacceptable. But for three engines, outage time would be 0.1 percent, or 8.8 hours/year, which is reasonable. This means:

- At least three engines must be on line at all times (the spinning reserve requirement) at two-thirds full load each, so that if one trips the remaining two will take a full load each to maintain the baseload output. If four engines are kept on line, each would have to be loaded to at least 50 percent of full load.

- At least one engine must be ready to be started when an outage trip occurs in an operating engine (a hot standby).

- At least one engine must be undergoing short-term (24-hour) maintenance (a cold standby).

- At least one engine must be undergoing long-term (1- to 2-month) overhaul (a spare).

- *For constant load, a minimum of six installed engines are needed to ensure less than 9 hours/year outage.*

Module sizing considerations

The previous analysis applies for base-load conditions only. Assuming a peak load twice the minimum load increases the on-line engine count to five and the installed count to eight. On average, four engines would be on line, or 50 percent of installed capacity.

It is very important that *all modules be the same size* and carry the same load when on line, otherwise they will not be able to back each other up. Similar-size modules make maintenance simpler too by reducing spare parts and tool inventory.

Other Cogen Plants' Reliability

Overall reliability is less for gas turbines and steam turbines. Start-up is more complicated for gas turbines and much longer for steam plants, therefore paralleling the plant with a utility grid becomes the only feasible reliability alternative.

Heat loads do not require the same degree of reliability as electric loads because it takes much longer to notice the loss of heat. This fact reduces backup requirements substantially for heat loads.

Gas Turbines

The great advantage of gas turbines is that they do not need cooling water and consequently can be installed in the middle of a desert! Although they do not have minimum load limitations, they are very inefficient to operate in the island mode since then they cannot always run at full load but must follow the available demand. Natural gas is the best fuel for a gas turbine.

Small Hydro Plants

Small hydro plants are reliable and inexpensive to operate. They are an excellent power and energy source when a water resource is available. A small dam will store water for peak load hours.

Wind Farms

Wind farms are erratic sources of energy. To tame wind energy, some practical form of energy storage must be provided, such as batteries. To interconnect a wind farm to a utility grid creates serious instability, especially of reactive power if the generators are the induction type.

Steam Turbines

There are three types of steam turbine installations: condensing, back-pressure, and extraction.

Condensing steam turbines are mostly in service with utilities in large power plants. The steam coming from the turbine is exhausted into an air- or water-cooled vacuum condenser. No waste heat is recovered for other useful processes, but instead is dissipated into the environment. Consequently, the thermal efficiency of such an installation is only about 35 percent.

Back-pressure (BP) steam turbines are used in many chemical and food plants. The turbine exhaust steam is used for the process, which acts as a virtual condenser. This scheme results in efficient energy utilization cycles.

BP steam turbine load will depend on the available process-steam load. *Therefore, these power plants cannot operate in the island mode unless there is enough standby electric generation for load cycling, and/or a heat dump or storage device, e.g., a condenser or a steam accumulator.*

Extraction steam turbines can be either condensing or back-pressure, but steam is also supplied by the intermediate stages at higher pressures than the final exhaust pressure. Steam does not necessarily have to be extracted from these stages, but can also be recovered from the process waste heat through an intermediate pressure drum and supplied back to the turbine for increased electric power output.

Combined-Cycle Cogeneration

Combined-cycle plants [gas turbine plus HRSG (waste-heat boiler) plus steam turbine] are flexible and efficient power plants. The gas turbine should run always at full load (approximately 70 percent of installed capacity) for most efficient operation. In addition, the steam turbine can be either condensing or back-pressure.

Load cycling can be handled by the steam turbine with exhaust duct burners to increase steam production and a diverter damper to dump some of the gas turbine exhaust heat to the atmosphere. Combustible by-products of the process plant may be used as fuel in the waste-heat boiler for supplemental firing.

Combined-cycle power plants are presently popular also with utilities all over the world because of their flexibility and relatively low-cost installation. Steam is discharged into a condenser instead of being used for a chemical process.

Cogeneration Steam Plant Pressure Control

Since there may easily be three different steam pressure headers—main, extraction, and exhaust—in a steam-driven cogeneration plant, steam pressure control becomes fairly involved. Each header may have more than one boiler or pressure-reducing station directly connected to it, each one with its own pressure-control valve.

Only one steam pressure signal and controller, per header, can be used for control purposes, even when there is more than one steam pressure control source on that header. If more than one pressure signal were used for pressure control on the same header, the differences in calibration and accuracy would create a conflict. Also, if more than one controller would try to maintain the pressure on that particular header, each controller would try to gain supremacy over the other, with the fastest and strongest pressure source locking out the rest.

It is possible to have one master controller, such as a load master, and a series of slaves to fine-tune the gain to the individual control sources. All the controllers but one should be *gain-only*, i.e., without integral action or reset, otherwise an unstable situation, as described above, will occur.

Since the different sources of steam pressure to the same header will have different characteristics, it is important to assign to each one a certain weight in its contribution to control and a set point. Assigning the set point actually amounts to introducing a split range; i.e., each source or control valve will start its contribution at a certain output from the load master controller.

It is of course possible to have all the pressure control valves to the same header respond simultaneously, but that is usually not required, since some sources are more economical and should be given preference. It is not clear that it is more economical to draw steam from a pressure-reducing station than to continuously fire a boiler up and down, but it certainly is more economical to draw steam from turbine exhaust or extractions.

Steam-pressure-reducing stations (SPRSs) are usually backup sources, since energy is lost in the reduction. To recklessly fire a boiler up or down also creates huge fuel losses because of imperfect furnace combustion conditions.

To capitalize on the advantages of both systems—boiler firing and pressure-reducing station—the response times of each source can be used to strike a compromise. Since boilers need considerable time to reestablish steam pressure and overfiring is uneconomical, the pressure-reducing station can be used to meet the short-term steam demand while the boiler is engaged only to *slowly* reestablish itself as the main steam source.

Steam-Pressure-Control Case History

A combined-cycle plant has a gas turbine with a supplementally gas-fired HRSG boiler with back-pressure and one extraction steam turbine (Fig. 6.1). Consequently, there are three steam headers: high-, medium-, and low-pressure.

Only medium- and low-pressure steam are used for process purposes. To supplement steam generation, there are three package boilers (PBs): two that supply the medium steam pressure header and one that supplies the low-pressure header.

The package boilers are water-tube boilers and *cannot* store steam in substantial amounts the way a fire-tube boiler would do (Fig. 6.2). In order to avoid excessive steam pressure drop in the drum, they are fitted with back-pressure-regulating valves! (Once again, low-pressure process steam for backup should come preferably from fire-tube boilers, which can absorb large fluctuations.)

Steam to the medium- and low-pressure headers is also supplied via pressure-reducing valves from the high-pressure header, for backup purposes.

Split-Range Control Action

The control philosophy should be gain only for SPRSs and integral action only for PBs, given that:

- Most of the steam is used by the low-pressure header.

- The main source of low-pressure steam is the steam-turbine exhaust.

- There is no substantial steam storage from a fire-tube boiler or other medium.

Under these circumstances,

- The waste heat from the gas turbine provides the minimum steam load heat, and additional demand is met through supplemental firing of burners installed in gas-turbine exhaust ducts and the package boilers.

- The steam turbogenerator runs flat out on the available steam pressure. The burners are controlled by the high-pressure steam header, to maintain the steam pressure.

- The intermediate- and low-steam-pressure headers maintain pressure via control of the SPRSs and the PBs.

In this way, as soon as the load master pressure controller detects a set-point deviation it sends a signal to the slave controllers. One slave will control the SRPSs and the other the PBs. Each station and each boiler may have its own slave and auxiliary controllers.

In order to keep the steam pressure from drifting excessively, the SPRSs compensate for the change in steam load. The PBs in the mean

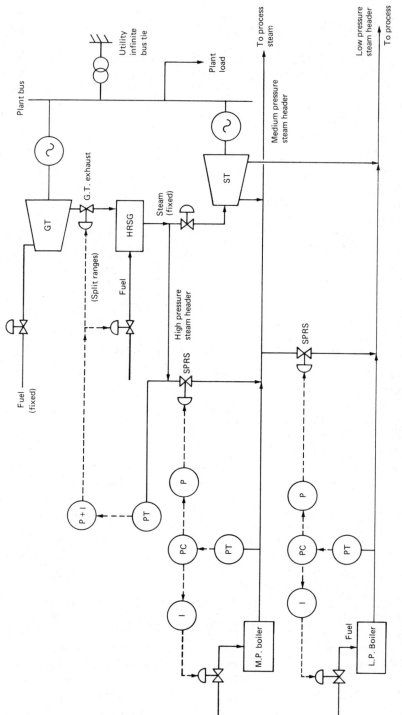

Figure 6.1 Combined-cycle cogen plant. P: gain; I: reset; PC: pressure controller; GT: gas-turbine; HRSG: heat recovery stream generator; MP: medium pressure; LP: low pressure; and PT: pressure transmitter.

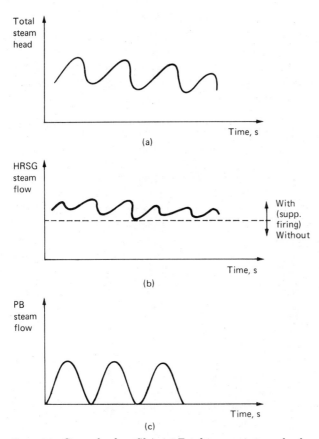

Figure 6.2 Steam load profiles. (*a*) Total (process) steam load. (*b*) HRSG steam load. (*c*) Package boiler steam load (*a* and *b*).

time begin to change their firing rate to *slowly* readjust for the new steam load. By changing the firing rate slowly, near-perfect burner and furnace conditions can be maintained and fuel losses reduced.

Making Interconnections Work[2]

The electric part of a cogeneration plant can be fairly simple since, *regardless which way the power flows, the same voltage, frequency, transformers, and switches will be used*. The real challenge consists in *controlling the flow of power*.

In the island mode, the engines' governors are at the heart of the control system. Droop control must be always in place, even when the system is connected to the utility's infinite bus, so as to protect against a sudden loss of tie load and consequent engine under/overspeed.

Electromechanical engine governors with mechanically adjustable speed/load and droop are preferred in the island mode since in case the electronic load/frequency controls fail, the plant can still operate, albeit under manual control. Excitation control will keep voltage within tolerance during major load upsets, including short circuits. *It is impossible to operate island-mode power plant generators without governor droop and voltage regulators.*

Whether your plant is isolated (island mode) or connected to the grid (infinite-bus mode), careful design and adjustment of the active (kW) and reactive (kVAR) load controllers are necessary for smooth operation. Power factor (PF) becomes particularly relevant since now the cogenerator, whether it wants it or not, has control over it.

It is very common—and wrong—to assume that power factor will take care of itself and therefore can be disregarded to the point of not even having an instrument to indicate PF. Synchronous generators control PF through excitation control, and *PF must be adjusted equally among all the generators and the utility tie.*

Operation of an isolated system is completely different from one that is connected to a grid. See Table 6.1.

Bus Voltage Selection

Voltage selection is based on the present and future load requirements. For cogeneration projects, it is preferable to select the existing installation's bus voltage in order to limit capital investment costs and keep the same transformers and circuit breakers. High-voltage

TABLE 6.1   Comparison of Cogeneration Plant for Island-Mode and Utility-Grid Operation

Plant characteristic	Island mode	On utility grid
Generator type	Synchronous	Induction preferred
Load-speed governor	Frequency control only	Power exchange with utility
Excitation control (synchronous generators only)	Voltage control, PF balance between generators	PF balance, kVAR exchange with utility
Engine start-up	Black-start capability	Utility-supplied power
Utility contract	Off-line standby where possible	Demand charges, energy charges, penalties
Installed capacity	Twice average electrical load (approximately)	Not to exceed maximum waste-heat load
Bus protection	Low voltage, low frequency	Inadvertent power flow

equipment is more expensive to install and maintain and requires skilled personnel and strictly enforced safety procedures.

High currents, on the other hand, also require expensive installations. Low-voltage bus breakers are available in the 4000 continuous amperes range, but this is probably the practical limit.

Generator Selection

Rotating ac generators can be either induction or synchronous machines. Usually induction generators are used with small prime movers connected to a grid (infinite bus), while synchronous generators are used with large prime movers and stand-alone (island-mode) operation.

Induction generators

Induction generators are induction motors which, instead of supplying torque, are being driven by the prime mover past their synchronous speed until they develop a braking torque and supply power into the grid. See the torque-speed graph in Fig. 6.3.

Induction generators need to be excited by the grid to which they are connected, so they are well-suited for small cogeneration projects. Frequency and voltage are fixed by the grid, and therefore there is no need of frequency and voltage controllers, with all their additional complications.

With a capacitor bank matching the induction machine's own reactance, the system can run in island-mode once excited. The bank of capacitors can be connected in parallel to diminish the reactive load they present. But a capacitor bank poses frequency- and voltage-control problems when the load fluctuates. In island mode, the "slip" speed of the prime mover controls the frequency, and the reactive load of the capacitor bank controls the voltage. Consequently, for an induction generator, prime-mover speed and capacitor-bank reactive load would have to be continuously readjusted to keep the frequency and voltage constant. This makes use of an induction generator impractical in island-mode operation.

Synchronous generators

It is simpler and more practical to control frequency and voltage with a synchronous generator. Frequency is controlled by maintaining constant speed (as opposed to varying speed for the induction generator) and by adequately opening the fuel or steam valves to the prime mover.

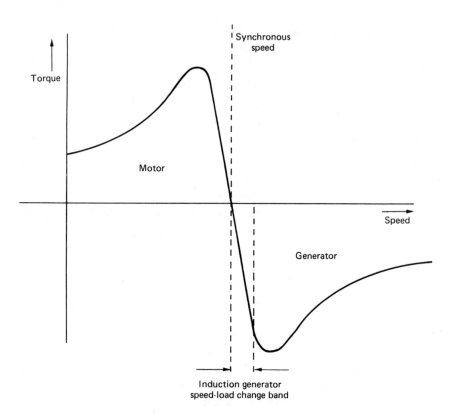

Figure 6.3   Induction machine torque-speed characteristic.

To control voltage in the island mode, the field excitation control is used. Field excitation control requires substantially less energy than line excitation control via a capacitor bank, as required for induction machines.

DC generators/AC inverters

It is possible to generate ac via dc with battery storage and solid-state or electromechanical dc-to-ac inverters. This method has many disadvantages:

- It is energy-inefficient.
- It is capital-intensive.
- It generates large amounts of harmonics with solid-state inverters.
- It generates large inductive loads.

Both the dc generator and the ac solid-state inverters have low efficiencies compared to rotating ac generators. With the exception of high-voltage dc (HVDC) transmission, dc/ac generation is used only when energy is free, as in solar and wind projects, or must be stored in batteries.

References

1. W. Goldstern, *Steam Storage Installations*, chap. 2, Pergamon Press, New York.
2. J. Reason, "Making Interconnections Work," *Power*, June 1982.

Power System Control and Operation

Basic Principles

The three functional requirements of an electrical power system are, in order of importance:

1. Reliability (service continuity)
2. Quality (voltage and frequency stability)
3. Economy (efficient planning and operation)

To satisfy these three requirements simultaneously is not a simple task.

The primary purpose of an electrical energy system (EES) is to deliver energy, not electricity. The electrical link of an EES is only the result of choosing a medium that is more convenient to handle for specific applications, as compared to other forms of energy distribution, say, fuel or steam. As a result, mechanical engineering actually takes precedence over electrical power engineering. In the same way, electrical power engineering takes precedence over electronic communications, instrumentation engineering, and computer science. Regardless of all the conveniences that a computerized operation can bring, it cannot replace a skilled crew of workers, which is absolutely necessary to operate, maintain, and repair a system and assure continuity of service.

An EES can be viewed as a gigantic power converter made up of various smaller links: alternators, transmission lines, transformers, all the way up to the low-voltage distribution network and final user.

The users invariably use electrical energy in applications that convert it into heat, and heat (kilowatthours) is what they get billed for. Ergo, we are delivering and selling heat.

An alternator is a mechanical-to-electrical power converter. A transformer is an electrical-to-electrical power converter. Each link in the chain, from the prime mover to the final user, has its peculiarities. Although there would not be any electrical system without an energy source, it would be impossible to generate it without a user, or demand for it, as well.

Control of Electrical Energy Systems

Demand by users is random, and when demand for active and reactive power changes, corresponding changes occur in frequency and voltage. Consequently, in order to control generation in a power system, it is necessary to have changes in voltage and frequency first. At the same time, the control equipment must keep these changes within bounds. A deficient response from generating equipment will produce unwanted and even intolerable changes in frequency and voltage.

That is why, even in very old equipment, it is possible to find speed, voltage, and steam boiler controllers. Reliable operation would be impossible without automatic controllers, and, as power plants and distribution areas become more and more interconnected, new requirements are imposed on controllers. Nowadays we talk about *hierarchical* (multilevel) control. We also are aware that control and stability go together.

Operating economy is also affected by the *dynamic* behavior, especially in fossil-fired steam plants, since efficiency diminishes with load changes. The more abrupt—i.e., faster—these changes are, the more efficiency is lost in the process.[1] As early as the planning stages, it is necessary to consider dynamic behavior of generating equipment and how it will affect the rest of the system's operation.

Hydroelectric plants are the best-suited and most efficient to satisfy load demand changes. But, of course, hydroelectric plants cannot be installed everywhere. Steam plants should stay at base load most of the time—the base-load boiler firing rate, that is. A steam turbine can and should be allowed to follow small load fluctuations, as long as it does not affect the average steam pressure significantly. Steam accumulation in boilers with high main time constants help to make steam pressure less susceptible to load changes. This improves dynamic behavior by damping out small erratic changes in load.

The following are some of the tasks to be performed by a power system control group:

- Frequency and stability-margin control in real time
- Equipment testing (basically alternator field controllers and speed regulators)

- Power imbalance relay coordination (low-frequency relays, power-flow relays)
- System modeling and identification
- Dynamic optimization, to improve stability and economy

Equipment Requirements

Controllers, in general, should not be more sophisticated than a specific plant requires.

Turbine speed controllers

Optimum droop will depend on the system configuration at any given time. Therefore, it is necessary to be able to change droop, over a sufficiently wide range, as many times as necessary, especially if underfrequency relays are in play. This can be effectively accomplished only by means of speed controllers with an electronic droop adjustment.

Interlocks must change the droop settings as soon as the system configuration, and ensuing droop requirements, change. Speed of response of the speed controller has to match dynamic characteristics of steam or water flow to the turbine, steam line, or forced tube, respectively.

Dead band becomes more obvious as more turbogenerators are paralleled. What used to be an acceptable margin, 0.06 percent, is no longer. Dead band is directly responsible for sustained frequency oscillations with periods of 10 s, and can be recognized by this feature.

Alternator field excitation controllers

These functions deserve special attention:

- Reactive load control to minimize circulation currents
- Accurate response for any load change condition
- Fast response, with positive oscillation damping during faults or upsets
- Limiting voltage surges when alternators trip

These functions also often deserve attention:

- Limiting δ angle
- Power-frequency (P-f) oscillation damping

Perhaps the most important function of the alternator field controller is reactive load compensation so that generators can run parallel on the same bus *at the same power factor*. Automatic reactive load compensation is hard to implement accurately enough that there will not be an accumulated drift and consequent power factor imbalance between alternators feeding the same bus. Stability problems, spontaneous oscillatory conditions before or after a fault, and even loss of synchronism originate in poor reactive load compensation, coordination, and controller response. After a fault, or a major load upset, large over- or underexcitation margins by both the field controller and the alternator field are highly desirable in order to stay synchronized and limit overvoltages. Positive damping of oscillations must be verified.

A feature of modern alternator field controller technology is protection to avoid excessive δ angle swings, in order to stay synchronized. However, overexcitation of alternator fields has little influence during faults because of the field's large time constant (~ 5 s) and its relative location in the control loop. The energy stored in the alternator's field is also a cause of overvoltages when there is a sudden loss of load because of a circuit-breaker trip.

This overvoltage creates an additional burden on any bus breaker because it increases the possibility of a restrike of the arc inside the breaker. An arc restrike can cause the breaker to explode, if it is of the oil type, and can cause other serious damage as well. To reduce alternator trip type overvoltage, modern field regulators can generate large inverse field currents to reduce the field much faster. Modern alternator field controllers also participate in *P-f* control, by exploiting the network's kilowatts-versus-volts characteristic and utilizing it for *P-f* oscillation damping, with mixed results.

Reliability

Electric power plants need intensive maintenance, and regardless of how efficiently they are managed, they will have outages, partial or total, scheduled or not. To minimize backup requirements, electric power plants are interconnected to power grids. Not only reliability, but also quality and economy benefit from this arrangement.

Power Plant Management

Contrary to common wisdom, the engineering and design of a power plant does not end at start-up but has to continue for the rest of the plant's useful life. Design engineering can provide only the minimum requirements for a succesful plant start-up and operation, because it is impossible to anticipate all the contingencies that will occur in the years to come. What looks good on a blueprint will not necessarily translate into equally good results. Recognition of this fact by management is crucial to good return on investment through reduced frequency and duration of outages and elimination of costly mishaps. There are no "accidents," only poor design, planning, and management.

Some of the most common sources of trouble—and consequently areas for improvement—are

- Negative attitudes and laissez-faire management policies
- Poor understanding of processes and systems
- Nuisance trips due to inconsistent logic, poor controller loop design and tuning, lack of false-alarm suppression, and insufficient operator training for emergency conditions

Unless a specific occurrence is new and undocumented, all incidents can be traced to poor management procedures.

Dynamics of Systems

In any system, demand is the cause of all that happens, the only independent variable. We can influence it, change it, create it, and destroy it, but without demand, we do not have a system. Everything else in a system is designed to satisfy demand with reliability, quality, and economy.

Demand can take many different forms. It can be constant (we wish it would always be!), but usually it is variable in time, and that is why problems start.

A variable demand causes serious stress in a system, because the system must be able to satisfy it in all its forms, and there are a lot of them. Demand may be (to name a few possibilities)

Seasonal

Global

Regional

Daily

Hourly

Weekly

Monthly

Yearly

Instantaneous

Growing

Shrinking

Minimum

Maximum

Average

Peak

Low

Position/step

Velocity/ramp

High

Valley

Erratic

Low-noise

Meeting all these different conditions at one time or another in a single system is a tall order. If we design a system to meet a certain maximum demand, we are assuming a certain economy based on this demand, but maximum demand may occur for only, say, a minute a year. The rest of the time we are saddled with an average demand of, say, 50 percent. Suppose an investment has to be made for 100 percent demand, but we have only a 50 percent average. What if we do not meet the other 50 percent? The whole system may collapse, as in the 1976 blackout of New York City.

Another, more insidious, type of demand, is the instantaneous kind. Regardless of what the average, or long-term, demand may be, there is an instantaneous demand that can change capriciously because of a variety of factors over which we normally have no control. We can only *be prepared* for when it occurs.

A system that we so enthusiastically designed and carefully groomed can suddenly beset us with all kinds of demands for which we are not prepared because we did not even know they existed. Suppose, for example, we decided to build a cogeneration plant because planners said it would be so good for our finances. Besides, we were told, when the utility goes down, *we* will still have power. We were shown wonderful calculations and forecasts of the money we were going to save. It looked good on paper. So many engineers could not be wrong. It was their job to know. But, as usual, engineers had to take a back seat behind business and politics, and the project failed financially.

Stability

Power system stability may take different forms, depending on the state of the network. During fault conditions, the primary concern is the ability of the network to stay interconnected. The loss of interconnections will only increase instability. During normal operation, the weight is shifted to keep voltage and frequency fluctuations small.

During fault conditions we must deal with *absolute stability*; i.e., can the network stay interconnected or not? During normal conditions we deal with *relative stability*, i.e., keeping voltage and frequency within an acceptable error band.

Network stability is heavily dependent on the load. Low loads during night hours are sometimes more difficult to deal with than peak load. Steam and diesel plants can reduce their loads to only about 50 percent of full load; below that, they begin to shut down. Daily start-up and shutdown cycles create tremendous thermal stress and reduction of useful life, especially in steam plants.

Lightly loaded power lines (especially underground cables) are like capacitors. They force leading power factors on generators, making them increasingly unstable, and may force reactive power "wheeling" with associated thermal losses.

On the other hand, heavy loads require huge amounts of spinning reserve to meet possible plant outages. Interconnections should never trip on overload, but should stay interconnected up to their thermal limit. Peak-load and base-load plants must be coordinated as one, on the assumption that they will trip together, so that the loss of either one will not affect the rest of the network.

But probably the grid's largest source of instability is poor power plant design, operation, and maintenance. Here are a few such items that affect power plant stability—all of them avoidable:

- Plant generators with uncoordinated voltage/power-factor regulators
- Locked steam-turbine governors
- Poor protective relaying design and coordination
- Voltage-dip-sensitive pumps and motor starters
- Voltage-dip-sensitive plant interlocks
- Poor PID steam boiler controller loop design and tuning
- Poor hydro plant governor calibration
- Poor start-up and shutdown procedures
- "Take-it-for-granted" attitudes
- Excessive simultaneous alarms
- Non-fail-safe design

Any one of these can shut down a plant for no good reason.

Uncoordinated voltage-regulator trips

Uncoordinated alternator voltage regulators will create contradictory responses during major network voltage upheavals, with eventual loss of synchronism of one or more generators. Both static response, i.e., slow reactive load distribution to maintain uniform power factor, and dynamic or fast fault-stabilizing response must be coordinated.

Trips due to nuisance valve and pump trips

It is common for power plants to shut down because of a pump or valve that tripped for no apparent reason. This usually can be traced to a voltage dip that caused a relay or a starter to deenergize. This is a ubiquitous type of problem which is time-consuming and sometimes expensive to correct. A stabilized power source must be provided to the coil that engages the breaker, starter, or contactor.

PID controller loop design and tuning

Poor PID controller loop design and tuning is a pervasive problem in the process and power industry. Even if the loop happens to be designed correctly, the controller settings will still have a high degree of probability of being set incorrectly. The drum level control loop is a typical case. A poorly designed and tuned drum level loop will force a boiler shutdown after a major load upset from an excessively high or low level. Load master (steam pressure) control loops are also nuisance trip targets, especially if there is more than one boiler on the same header.

Plant operators

A major plant control element is the human operator. As with "accidents," there are no "operator errors," however, only design, planning, and management errors. It is wrong to assume that it is an operator's duty to become "Sherlock Holmes." There is too much at stake. Some types of errors are not supposed to happen *ever*, e.g., meltdowns, furnace explosions, or generator breakers closing out-of-synch. In order for the operator to function as efficiently as possible, only selected pieces of information should be presented during emergency conditions. Alarms must be selectively "blacked out," interlocked, and presented on a priority basis. Abnormal conditions are, obviously, not frequent, so there is little chance to gain familiarity or provide complete training. Consequently, the exact source and explicit solution of the problem must be presented to the operator. It is unrealistic otherwise to expect the operator to make an accurate assessment of the situation under stressful conditions from excessive data.

Fail-safe design

There should be a law requiring all industrial plants to be designed fail-safe; that is, to be designed so that, no matter what the *single* occurrence, the plant will shut down safely. Typical examples of non-fail-safe designs are most nuclear reactors and automobiles. Airplanes are designed fail-safe because they have built-in redundancies. Most nuclear reactors are not fail-safe by definition because a total loss of power would cause a meltdown. Obviously there is an element of subjectivity here, because we may accept the fact that planes crash but not that nuclear power plants "melt."

Control wiring for alarm signal loops fail safe in the energized position; that is, they provide an alarm when electrical continuity is lost. Equipment fails safe in the electrically open position, so that when control power is lost the equipment stops by itself.

System Stability States

Table 7.1 provides an overview of the three different states of a power system. Obviously static stability does not matter to the system, since normal operation is too far from those limits. The table gives only a qualitative indication of the stability margins.

TABLE 7.1   Stability States of a Power System

	State of system		
	Fault	Normal	Permanent
Type of stability	Absolute	Relative	Static
Control type	Reclosing	Primary/governor	Load flow
Control means	Breakers/valves Power factor*	P-f	P and Q
Disturbance magnitude	> 10%	< 10%	0
Disturbance duration	1 s	< 300 s	∞
Error signal	Current $I(t)$	ΔP, Δf	P, Q
Type of analysis	Nonlinear	Linear	Economic
State variables	$\dot{x} = f(x,u,p)$	$\dot{x} = Ax + Bu + Cp$	$0 = f(x, u, p)$

P = power, f = frequency or function, Q = flow, V = voltage, ΔP = power drift, Δf = frequency drift, x = state variable, u = control variable, p = disturbance variable.
*Effective only prior to fault.

Reference

1. G. Quazza, "Role of Power System Control in Overall System Operation," *Brown Boveri 1971 Symposium Proceedings*, E. Handschin (ed.), Elsevier, New York, 1972.

Chapter 8

Power-Frequency Control in Electrical Energy Systems

Power-frequency control is first of all a mechanical, not an electrical, problem. Regardless how small or how large, electrical energy system prime movers must adapt continuously to demand. The instantaneous difference between generation and demand translates into small, incremental speed changes, ergo frequency changes for *all* the generators and generating plants synchronized to the same grid (Fig. 8.1). Since a change in demand always anticipates the generating effort, there is a continuous change in the system's frequency. Frequency, being so sensitive to load/generation imbalances, gives a strong indication of the quality and, to a lesser extent, the reliability of the network. A very stable frequency indicates a very responsive grid; consequently, the grid must be reliable. An excessively fluctuating frequency indicates a "chaotic" grid, with poor coordination among generating plants; i.e., it is unreliable. Frequency fluctuations with typical 10-s periods are an indication of excessive governor dead band (see "Turbine Speed Controllers," Chap. 7.). Curiously enough, just by recording and monitoring frequency from a 110-V outlet anywhere, one can infer a lot of what is going on in the power pool, since the frequency is uniform in a synchronous grid.

Load Response to Changes in Frequency

Since most electromechanical loads are also inductive, it is only natural to assume that when frequency increases, active load will decrease, and vice versa. This is not the case at all in electrical energy systems. Most of the load is made up of electric motors, and electric motors will run faster when frequency increases, and therefore will pick up more load accordingly. The real situation in a power grid is

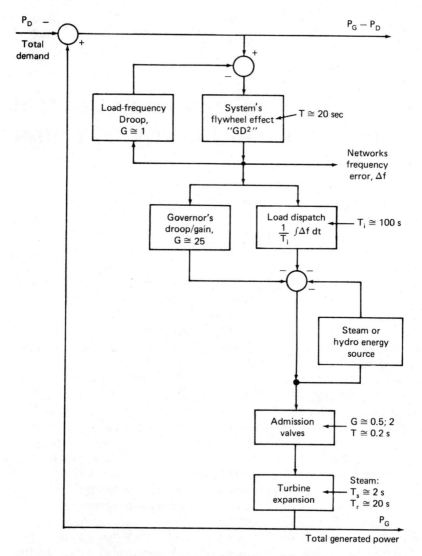

Figure 8.1 Power-frequency control block diagram model of a power pool.

that load increases with frequency, and vice versa. This property is called the grid's *stiffness, droop*, or *frequency characteristic*. It is a welcome effect because it adds stability to the grid by creating self-regulation, or negative feedback.

Speed Governors and Droop

Speed governors are the backbone of power-frequency control (**Fig. 8.2**). Governors are primarily in place to protect turbines from

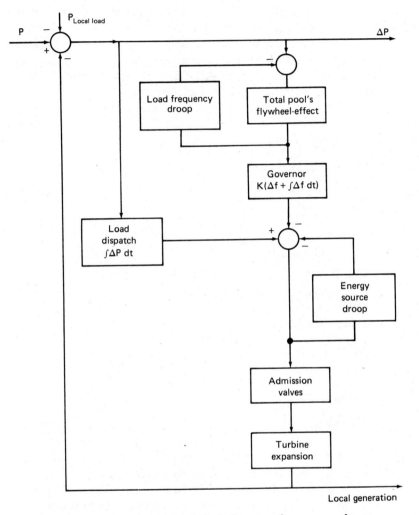

Figure 8.2   Power-frequency control block diagram of a power pool area.

overspeeding, without shutting them down. Governors can also be construed to be *negative-feedback amplifiers*. Governor *droop* is defined as the total speed change between no load and full load. When it is expressed in percent, it is the same as proportional band in process control. The inverse of droop (proportional band) is gain. For example, a 4 percent droop is equivalent to a gain of 25.

Since the governor only translates rotating speed into a valve position, it happens that *real droop* fluctuates according to process conditions (e.g., inlet pressure) and the opening characteristic of the admission control valves. Large steam turbo generators have many admission control valves, all of them operating in coordination. But it is hard to obtain a linear response, and real droop is therefore not constant but varies considerably. In order to keep droop within acceptable margins, special tests must be conducted periodically, according to standards developed by, for example, ASME (American Society of Mechanical Engineers) PTC 20, or other qualified bodies.

Governors used to be mechanical-hydraulic and were very reliable. Modern turbine governors are usually electrical and incorporate a multitude of additional features, like

- Automatic control of turbine load
- First-stage pressure control
- Valve testing
- Overspeed testing
- High and low output limits
- Valve position limits
- Automatic frequency participation

They also provide interfaces to

- Remote load dispatch
- Synchronizer
- Boiler control systems

Power systems keep their frequency deviations during normal operation *below 0.1 percent*! This is an extraordinary engineering feat, considering that systems can have more than a thousand generators on line at one time. On the other hand, because of the random nature of disturbances, the network also benefits from its large size, since there is a statistical tendency for disturbances to average out through the interconnecting transmission lines.

If, because of disturbances and the affected power plants' corrective action, there is a steady imbalance in the exchange of energy previously agreed on between areas, then *slow corrective action is taken by the load dispatch of the deficient area's* generating units to correct for the deficit.

Note that *it is the undergenerating plant that has to initiate corrective action* in order to restore the network to its previously agreed exchange schedules. If it were otherwise, the whole power pool might collapse, since all plants would not stubbornly stick to their schedules, thereby forcing the frequency up or down.

This also suggests plant size as a major concern. Indeed, any bus or node that carries more than the spinning (i.e., immediately available) reserve capacity of the total area load may well be responsible for extensive blackouts, since a fault at such a bus or node means that the rest of the area must make up for that deficiency. Since any bus or node can be at fault at any time, statistically sufficient spinning reserve generating capacity must be available at all times to accommodate such faults.

This works both ways: too much spinning reserve may be a problem when thermal units are in place, since there is a low limit to what these can carry, roughly 30 to 50 percent of full load. Small networks and island-mode plants are especially vulnerable, since there is an economic limitation to the degree of reliability that can be built into a stand-alone power system.

Interconnecting power plants increase overall reliability. Therefore, the larger and more interconnected the grid, the more reliable it becomes. System interconnection is the best method to deal with load fluctuations, since there is a tendency for the load to average out statistically.

Nevertheless, island-mode operation must still be available as a fallback survival option if large portions of generating capacity become suddenly unavailable. (This is what should have happened during the New York blackout of 1976.) If large portions of generation become suddenly unavailable, and generation can no longer satisfy demand, it is essential for the remaining generation plants to "cut their losses" in order to survive a total blackout and be able to restore service as soon as generation becomes available again. This is all too well known in places with precarious electrical energy systems.

Otherwise, when a total blackout occurs, there is no power to restart any portion of the system, and smaller emergency power units have to be pressed into service to start a slow and progressive buildup.

Take for example the following system case study:

1. Two oil/gas-fired steam plants, one with roughly twice the capacity of the other, are interconnected by a transmission line.

2. Periodically, a substation line-fault trip occurs, elsewhere in the interconnected system, and 20 to 40 percent of the *total load* is dropped.

3. Now, it happens that the larger plant has an automatic secondary load control loop, prematurely installed for future use when the system is connected to a much larger power pool, to keep the generator's output constant. The smaller plant has only primary governor frequency control, at 4 percent droop, including network load-frequency response.

What are the consequences? Obviously frequency will rise very fast when a line fault occurs, and since the only *frequency regulation* in place is in the *smaller* plant, it will dump any excess load it can. Since the smaller plant has to absorb its own share *plus* the large plant's, it will drop 20 percent of its load plus 2 × 20 percent of the larger plant's load for a total system load drop of 60 percent. This is a tremendous load upset for any plant. Most plants cannot operate continuously below 50 percent maximum generating capacity.

Meanwhile, what has happened to the frequency? A 60 percent load drop, at 4 percent droop, yields a frequency error of

$$60\% \times 4\% = 2.4\%$$

At 60 Hz, this is an error of 1.44 Hz. This is enough to create serious problems for large computers and other sophisticated equipment that rely on a stable frequency to perform satisfactorily.

When the total load shed exceeds 33 percent of full load, then the entire smaller plant would trip on reverse power! At the larger plant, units could overspeed and kick out, maybe forcing a total blackout. Even if there is enough power available through the interconnection, the entire plant would then have to be restarted, and it can take several hours for normal boiler conditions to be reestablished.

Ripple Control

Instantaneous changes in frequency are called *ripple*. They are below 0.1 Hz, sometimes 0.06 Hz. They have to be met on an as-needed basis. This is usually done automatically by *secondary control* (*primary control* is the action of the governors on the turbine control valves). Not all units participate in *P-f* ripple control, only the largest and best-suited ones. For example, there is no point for a small unit, like a cogeneration plant, to participate in the effort to maintain the frequency, since its contribution would be practically unnoticed because of the size of the grid. On the other hand, very large units must participate in order to reduce the burden on the rest.

It is easy to adopt a philosophy of "Nobody is looking, so I do not contribute," since *P-f* control imposes an additional burden on those who choose to participate. Therefore, careful planning and procedures must be in place in order to provide enough *regulating power capacity* to the grid. Since demand can adopt many different forms, the right response capacity, and strategy, must be *in place before a specific situation occurs*. All possible scenarios must be analyzed and accounted for.

One way to look at demand is by its rate of change. During normal operating conditions, there is an inverse relation between disturbance magnitude and rate of change; i.e., the faster the change, the smaller the disturbance. To compensate for these small changes, or disturbances, a variety of tools are available, depending on the type of power plant and units in service. For example, hydro units are especially good for P-f regulation and control because they can pick up and shed load from 0 to 100 percent of capacity in a relatively short time. Within hydro units, the penstock, water hammer, will impose steep rate-of-change limitations. Provisions for spilling water to avoid downstream erosion damage also impose limitations that force slower rates of change.

Fossil-fired steam power plants do not have the same flexibility that hydro plants have. Some units cannot properly operate at less than 50 percent load. Gradual rates of change cannot exceed 2 to 5 percent per minute for the largest units. Step changes may be somewhat larger, up to 10 percent. Beyond these limits, fault conditions, severe thermal stress, and thermodynamic cycle imbalances may result in an outage.

Additional limitations occur for grated coal-fired and other solid-fuel-fired plants. They require substantial time lags for furnace conditions to stabilize.

Efficiency always suffers when fossil-fired plants must change load. But because of the compressible nature of steam, they have a load step-change capability that hydro plants do not. Steam plants can easily compensate for small pulse-type changes in the load, which average out after a while and do not require a change in the firing rate. This stored-heat capacity of steam plants should be exploited by load dispatch strategies.

Nuclear plants, on the other hand, come in two kinds: boiling-water reactors have one thermodynamic loop; pressurized-water reactors have two. Consequently, boiling-water reactors are much faster to respond.

To complicate matters even further, there is the question of efficiency. Hydro plants want to use all of their water, at the highest rate-hours in the market. Steam plants want to generate at their most efficient load. And most want to stay at a bottom of the demand curve, i.e., at base load.

To overcome reluctance to participate in regulating-type operation, those plants that pick up the burden to make up for load and frequency fluctuations should be allowed to charge extra for the service.

Keep in mind that typical instantaneous fluctuations of 1 or 2 percent of total area load will be a considerable amount of power for any single plant to pick up. For a plant to be able to participate in P-f control, its equipment must be especially responsive and stable. Boiler controls must be well-designed and tuned. Interlocks must not initiate nuisance trips because of excessive sensitivity to fault clearing, a common malady in power plants. Governor action must be fast and stable and must damp power fluctuations fast. Alternator field control also must be designed to damp fluctuations, of both active and reactive power.

Cogeneration plants may use diesel engines, gas turbines, or wind, solar, or geothermal energy, etc. They are relatively small, and therefore do not normally participate in *P-f* control in large interconnected systems, but they tend to deteriorate relative stability because they push up the base generation level. Small power plants are usually run at full load when they are on an infinite bus.

Diesel plants are very reliable and excellent for stand-alone, island-mode, or emergency power. They have the fastest start-up of any plant. In the island-mode, *P-f* control becomes just a matter of meeting two contradictory requirements:

1. Having enough spinning reserve to sustain at least one engine trip

2. Having enough engines on line without operating them below their low limit

Satisfying these requirements calls for plants with many but small modules. See Table 8.1 for an overview.

Automatic Power-Frequency Control

The first priority of automatic load dispatch is to maintain tie-line exchange schedules. Since load is constantly changing, either in or out of the controlled area, a constant turbine redispatch is necessary. This can only be executed by the *smaller area*, especially if it is importing power, otherwise the whole system would collapse!

Primary control, being a stabilizing action only, cannot handle the resulting steady-state error, while secondary control implies integrating the frequency error over time.

The error signal to a PID (i.e., analog) controller is

$$\text{Error} = \Delta P_i + K\,\Delta f$$

where K is the relative weight of the particular plant on the network's total regulating power capacity. Since it is also necessary to differentiate between normal and emergency network conditions, K will adaptively change accordingly. During normal conditions only selected plants participate in frequency control. When serious loss of load or generation occurs, it is imperative that all plants contribute. Plants can have a low K during normal network operation, and an increasingly higher K as conditions deteriorate for whatever reason.

All plants participating in secondary control must initiate corrective action when the frequency error exceeds a predetermined margin.

TABLE 8.1 Dynamic Classification of Power Plant Operation

Plant type	Remarks	Load disturbance			Reliability, availability‡	Cycling duty		Load economy			Plant load type
		Step*	Ramp	Pulse†		Loading flexibility	Start-up flexibility	50% load	(Ripple control) 50–100% load	100% load	
Hydroelectric											
High head	With compensation dam	—	2%/s	—	Peak load only	100%	High	Best	Best	Best	Peak
Medium head	With compensation dam	—	5%/s	—	High	100%	High	Best	Best	Best	Base/peak
Low head	Run of the river	—	10%/s	—	High	100%	High	Best	Best	Best	Base
Pump and generate	Thermal base-load cost	—	2%/s	—	High	100%	High	Very good	Very good	Very good	Peak
Nuclear reactor											
Pressurized water		—	1%/min	—	Low	Low	Low	Very good	NP	Very good	Base
Boiling water		10%	5%/min	200–400	Low	Good	Average	Very good	Very good	Very good	Base/peak
Boiler plants											
Coal	Natural circulation	5%	2%/min	200–400	Average	Low	Low	Very good	NP	Very good	Base
Oil/gas	Natural circulation	5%	5%/min	200–400	Average	Good	Average	Average	Average	Good	Peak/base
Combined cycle	Gas/steam turbine	5%	75%/s§	50	High	100%	High	Poor	Average	Good	Peak/base
Combined cycle	Supplemental firing	5%	65%/s§	50	High	100%	High	Average	Good	Very good	Peak/base
Once-through	Forced circulation	1%	1%/s	50	Average	Good	Average	Good	Good	Good	Peak/base
Back pressure	Cogeneration	5%	5%/min	200–400	Average	80%	Average	Good	Good	Good	Base
Sliding pressure	Natural/forced circulation	—	—	50–400	Average	50%	Average	Good	Good	Good	Base
Resource recovery	Solid waste fuel	5%	—	200–400	Average	NP	Low	Very good	Non-participant	Very good	Base
Internal combustion											
Diesel		100%	100%/s	—	High	50%	High	Poor	Poor	Average	Peak
Gas turbine		100%	100%/s	—	High	100%	High	Worst	Poor	Average	Peak
Cogeneration	with HRSG	100%	100%/s	—	High	50–100%	High	Average	Non-participant	Good	Peak/base

*Response at constant steam pressure.
†Boiler time constant, seconds.
‡Approximate yearly availability/reliability estimats. High: 90%, average: 80%, low: 70%.
NP: Nonparticipant
HRSG: Heat recovery steam generator
§Gas turbine only.

This action can be independent for each plant but must be coordinated overall. It must be slow enough that instabilities will not occur, but not so slow that the frequency will drift unacceptably.

Each plant will contribute according to its spare capacity and allowable rate of change. Consequently, some plants will respond faster than others, depending on the type of disturbance. Steam plants will respond better to step-load changes, while hydro plants will respond better to ramp load changes. Still, the differences in speed of response, especially between hydro and steam plants, are known to create instability situations that recur with long periods. These have to be addressed and proper compensation applied. Excitation control is one of the compensation methods currently used, with mixed results.

P-f control then consists of

1. Maintaining tie-line power exchange schedules

2. Eliminating frequency error

3. Dispatching load economically among plants

Automatic load dispatch must therefore be executed at both local and central levels. *Since economic dispatch has the lowest priority, it can happen that it will never take place, for reasons of network reliability and stability.*

Other Droops

For a power system to be stable, enough negative feedback must be built into it. They are different forms of droop, each with its own gain:

Type of droop	Gain, per unit
Governor	10 to 35
Load-frequency	0.6 to 1.5
Turbine torque-speed	Not available
Steam turbine inlet pressure	2.7
Load-voltage	0.6 to 2
Hydro plant governor	Various

The table shows the contributions of components of the system to its stability. The gains are meant for comparative purposes only. The governor provides the highest degree of stabilization, provided that the rest of the system follows suit. Steam generators (boilers) also provide stability because they supply additional power, like a loaded spring. Load diminishes together with voltage and frequency, providing an additional measure of stabilization. All turbines must be designed so that their torque *increases* when frequency falls, otherwise they would create a potentially dangerous instability. Hydro plants must have unstable configurations, that is, positive feedback loops, for protection purposes.

Types of System Mechanical-Load Disturbances

Typical load disturbances can be classified as steps, ramps, or pulses. Cyclic loads can be equated to a combination of positive and negative pulses. Pulse-type loads are applicable only to steam plants, where firing stays constant and steam pressure fluctuates up to about 5 percent. Step loads can be handled by some steam plants and internal combustion engines, but not by hydro plants. Ramp loads can be handled by all plants; internal combustion plants are fastest, followed by hydro, with steam plants the slowest of all.

9

Power Plant
Dynamic Control

Dynamic Efficiency

We saw in Chap. 4, "Dynamic Energy Control and Optimization," how efficiency is affected by control strategies. Whether energy efficiency in control loops becomes an issue depends on the application. For example, level control is not an energy-intensive application, but steam pressure control is, and so is any other application in which fuel or energy is the control variable.

Power plant managers know that plant operation will be more efficient if the load can be kept constant. This is a big load-dispatch issue for electric power plants connected to a power pool. Not only does net efficiency decrease, but equipment operation and maintenance becomes crucial in a power plant that is able to respond adequately to the dispatcher's requirements. Remember that network frequency fluctuations are the direct result of imbalances between generation and demand. It is the lack of generating response that causes direct fluctuations of the network's frequency, lowering dynamic stability overall. A premium rate therefore ought to be allocated to power plants with generating units that follow demand curves, based on kilowatt rate of change. If power plants have an incentive to follow the demand curve, the whole network benefits because reliability is increased through speed of response. If all the power plants on the grid make an effort to compensate for changes in demand, the frequency becomes that much more stable, and this is a clear dynamic stability improvement indicator.

Hydroelectric power plants are especially well-suited to handling large and fast ramp load change requirements. However, because of water-hammer limitations in the forced tube, they respond poorly to step load changes.

Fossil-fueled plants handle step loads up to 10 percent well and are especially adequate for load swings that do not require changes in the firing rate—i.e., swings with 5-min average cycle periods—since that is their stored thermal energy time constant. Solid-fueled plants in particular—i.e., grate-fed coal and refuse—cannot change their firing rate just to accommodate short-term load dispatch requirements. Pulverized-coal-, oil-, and gas-fired plants can, at rates of 5 percent/min, at the expense of becoming very inefficient. Figure 9.1 shows fuel cost efficiency for several types of power plant at different loads.

Pressurized water reactor (PWR) nuclear plants do not participate in load control because of the thermal stress it would cause in the fuel rods, but boiling water reactors (BWR) can respond very fast to demand changes from the load dispatcher. The rate-of-change limitations of steam plants in general derive from the steam generators and not from the turbine.

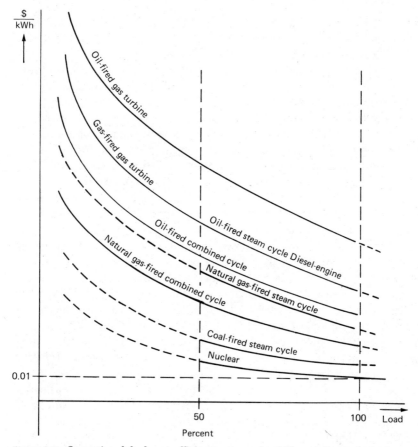

Figure 9.1 Operational fuel cost-efficiency curves for different thermal plants.

Gas turbines are very flexible to load changes, but at the cost of becoming very inefficient very fast. A gas turbine running at no load consumes fuel at a rate of 40 percent of full load! That is why gas turbines are used only as stand-alone prime movers for emergency and peaking power requirements, or, with waste-heat boilers, to generate steam for cogeneration and combined-cycle plants at full load.

Diesel prime movers are small compared to gas and steam turbines but offer distinctive advantages: they are easier and faster to start than any turbine and are the most reliable choice after stationary batteries and when large amounts of emergency power are needed.

Hydroelectric Power Plants

Of all the sources of energy presently available, hydroelectric is the cleanest, simplest, cheapest, and least threatening to the environment. This would be a wonderful world if all energy needs could be derived from hydroelectric power plants, but hydroelectric resources are limited both geographically and seasonally.

Hydro power plants provide primarily valuable flood and irrigation control; electric energy is only a by-product. There are environmental limitations to hydro power plants, but these can be accommodated to satisfy all the parties involved, and with careful planning, the long-term effects can be minimized.

All hydroelectric plants have their idiosyncrasies, depending on the characteristics of the hydraulic resource. There are four basic hydro power plant types:

- High head, or "peaking"
- Medium head
- Low head, run of the river
- Pumping and generating

In general, hydro plants are very flexible to schedule from a dispatcher's point-of-view—that is, if they include a compensating dam downstream to avoid erosion and flooding. They can be brought into service and meet changes in load demand from 0 to 100 percent within a minute. In contrast, they cannot meet step load changes because of hydraulic limitations such as water hammer and penstock resonance in the forced tube.

Since hydroelectric energy is inexpensive but limited, it is the goal of the dispatcher to maximize its utilization. Hydro plants are therefore sometimes in basic conflict with thermal power plants, which are also competing for preferential treatment from the dispatch center.

Hydrothermal dispatch, as it is called, results in complicated computer programs to accommodate daily and seasonal limitations of both thermal and hydro plants. Hydroelectric energy is usually billed at the "avoided cost" of running alternative sources of power. The major advantage of hydro plants is their absence of short-term (such as start-up) cost constraints.

For power-frequency control, high-head plants have the lowest load pickup rates at no more than 2 to 3 percent, still far better than any steam plant. Unloading can be much faster at the expense of wasting some water. They are ideally suited to be a standby reserve, much like a gas turbine, since the water reservoir usually allows only a few hours of daily operation.

For lower heads, the average daily availability of hydro power plants increases up to 24-hour run-of-the-river conditions, which change only seasonally. Low-head plants are well-suited for centralized power-frequency control, but run-of-the-river plants require that they be used at their maximum possible output which allows for short-cycle power-frequency stabilization only (i.e., ripple control).

Pumping or pumping-and-generating power plants are basically energy storage plants, much like rechargeable batteries. They are used to help flatten the daily demand curve. There are similar projects underway for compressed-air plants utilizing rock caverns as an energy reservoir.

Pressurized Water Reactor (PWR) Power Plants

Although nuclear power plants are well-suited for power-frequency control, PWR plants are kept at base load to avoid thermal stress in the fuel rods. The reasons for this become obvious from the locations of the PWR plant time constants in the control loop (Fig. 9.2).

Unlike a steam boiler, a PWR has its largest time constant *after* the load controller. It therefore requires severe control efforts to achieve the same speed of response as a steam boiler. This results in tremendous temperature buildup on the fuel rods, since the rate of steam generation in a PWR will depend on the temperature of the water in the primary loop.

Given the relatively large volume of water in the reactor, high rates of energy release are required to achieve only moderate rates of increase in the average primary water temperature. If the temperature becomes excessive, steam builds up around the fuel rods, impairing their cooling, a dangerous condition also known as *departure from nucleus boiling* (DNB). Similarly, to reduce output, stored energy (heat) cannot be dissipated too fast unless additional emergency cooling sources are brought into play.

The reactor's steam generator (the equivalent of the evaporator in a

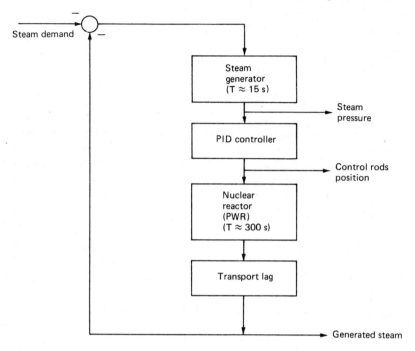

Figure 9.2   Pressurized water reactor steam pressure control block diagram.

boiler) has a time constant equal to that of the furnace of a boiler, while the reactor itself has a time constant equal to that of the evaporator, approximately 15 and 300 s respectively. Consequently, although boiler and PWR are dynamically similar, the *order* of the time constants is reversed and the beneficial effect of a high preceding time constant is lost.

PWRs also have a *transport lag*, or dead-time time constant between the reactor and the steam generator, which adds to their overall sluggishness and instability.

To avoid thermal stress in the fuel rods, a time constant can be created at the input of the controller of at least the same order of magnitude as the reactor's. The net effect is a very slow response to load changes required by the dispatch. With this additional limitation, PWRs become one of the most rigid type of plants in a power pool, making their use strictly for base load. This reduces the intrinsic value of their contribution. The steam generator's low time constant indicates a low steam storage capacity, making it unsuitable to respond to zero-average load fluctuations as well. PWRs and the power pool would benefit tremendously from larger-time-constant steam generators, i.e., higher steam storage capacity.

Boiling Water Reactor (BWR) Power Plants

Unlike PWRs, BWRs are very flexible in meeting the requirements of load dispatch, both for step and ramp loads, because steam is generated in the reactor itself. The time-constant location is the same as in a steam boiler. In addition, BWRs do not have the limitations boilers have in regard to air/fuel ratios.

The equivalent of a boiler's furnace heat-exchange time constant may be smaller than for steam boilers, since heat-transfer space is so much smaller. This also increases the reactor's response rate to load ramps, although this will ultimately depend on overall evaporating surface and steam temperature. The equivalent of the evaporator time constant will depend on the reactor's relative water content, just as in any other steam boiler.

The drawback of BWRs is that irradiated steam is used to drive the turbine, imposing serious containment requirements and increasing cleanup costs, already very high for any nuclear plant.

Curves for Dynamic Response of Steam Plants

Constant-steam-pressure power plant

Most power plants operate in the constant-steam-pressure mode, which provides the fastest response to load changes. The turbine control valves must provide additional pressure drop to control the flow of steam, thereby reducing the efficiency of the cycle.

The dynamic load response of this type of plant can be derived from the arrangement in Fig. 9.3. (Governor and turbine steam expansion time constants are assumed small.) As the figure shows, torque delivered by the turbine will depend primarily on steam pressure accumulated in

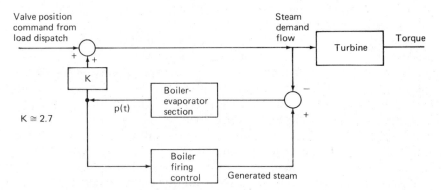

Figure 9.3 Block diagram for constant-steam-pressure/boiler-following power plant.

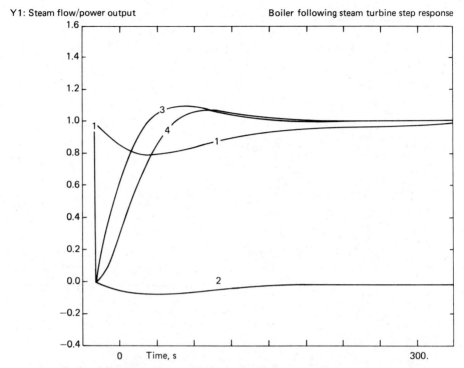

Figure 9.4 Boiler-following steam turbine step response. 1: steam flow/power output response; 2: steam pressure response; 3: fuel valve response; 4: generated steam response. Evaporator/main time constant is 300 s; furnace time constant is 15 s (i.e., no air-before-fuel). Boiler Load Master PID controller settings: gain = 10; integral action = 100 s; steam turbine flow/pressure ratio = 2.7/1.

the boiler. Turbine torque, i.e., steam flow, is very sensitive to changes in steam pressure (about 2.7 to 1). The delivered power transfer function is

$$\frac{P_G(s)}{\theta(s)} = \frac{s^3 + s^2/T_f + G/T_fT + G/T_iT_fT}{s^3 + s^2 \dfrac{T + T_f}{T_fT} + s\left(\dfrac{1 + G}{T_fT}\right) + \dfrac{G}{T_iT_fT}}$$

where G = control loop gain of boiler
    T_i = integral action of boiler
    T_f = furnace time constant
    T = evaporator time constant
    θ = admission valves position

Note that this particular arrangement does not include air-before-fuel control, which would increase the order of the model by at least 1, and slow it down. This arrangement gives close to the fastest response a steam turbine unit can provide.

The dynamic response curves in Figs. 9.4 and 9.5 show how unit response is affected by different boiler control parameters.

Y1: Steam flow/power output   Once-through boiler step response

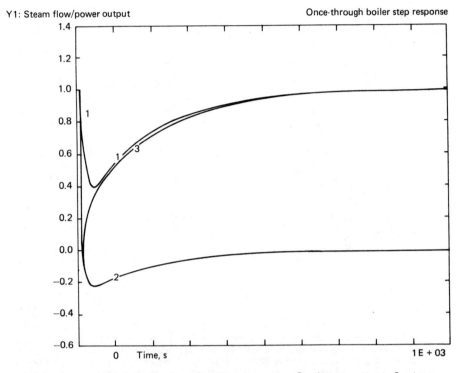

Figure 9.5   Once-through boiler step response. 1: steam flow/power output; 2: steam pressure; 3: fuel valve response. Main time constant is 50 s; furnace time constant is 30 s (i.e. air-before-fuel). Boiler Load Master PID controller settings: gain = 1; integral action = 50 s; steam turbine flow/pressure ratio = 2.7/1.

Sliding-steam-pressure power plant

Some power plants operate with boilers providing a variable steam pressure (Fig. 9.6). The turbine valves are completely open and provide only overspeed protection. The advantage of this type of operation is higher efficiency because of reduced pressure drop at the turbine throttling inlet valves.

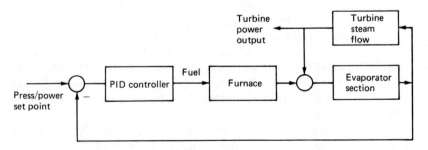

Figure 9.6   Sliding-pressure/turbine-following mode block diagram.

In order to provide as fast a response as possible, once-through steam boilers with very low evaporator time constants (about 50 s) are used. Power output is directly proportional to steam pressure. Response (Fig. 9.7) is excellent because of the low time constants involved.

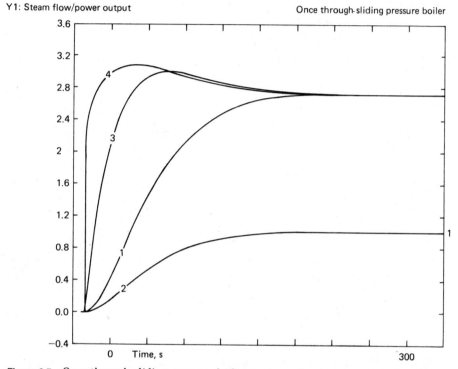

Y1: Steam flow/power output   Once through-sliding pressure boiler

Figure 9.7   Once-through sliding pressure boiler. 1: steam flow/power output; 2: steam pressure unit step response; 3: generated steam response; 4: fuel valve response. Evaporator/main time constant = 15 s; boiler Load Master PID controller settings: gain = 2.4; integral action = 48 s.

Chapter 10

Reactive Power-Voltage Control

Reactive Power-Voltage Control in Electrical Energy Systems

Reactive power can be described as a by-product of electrical energy systems. It circulates through generators, transmission lines, and transformers, but it is not delivered anywhere. It must be recognized and accounted for, however, since it plays an important role in the *stability and voltage control* of a system.

Reactive power is measured much like active power, that is, through a vector product. When we measure electric current alone, *we cannot recognize the active and reactive components of the current.*

Reactive power does not dissipate any energy other than the losses it creates through current circulation in equipment and lines. Every kind of load except a perfect heat load generates reactive power.

Since reactive power is either *leading* (capacitive Q_c) or *lagging* (inductive Q_L), the total balance must be zero. Alternators have the ability to generate both leading and lagging reactive power, but their ability to control it is hampered by stability considerations. Transformers generate lagging reactive power. Transmission lines generate either, depending on the load they carry. When a transmission line is lightly loaded, Q_c is larger than Q_L. Underground transmission cables are large sources of Q_c. When the network is lightly loaded, as it is at night, excessive Q_c becomes suddenly available, and this creates voltage control and stability problems.

In order to compensate for excessive Q_c, inductive reactors may have to be switched on and idle transmission lines disconnected. When everything else fails, recirculating ("wheeling") current through long transmission lines is used to help bring down excessive Q_c (Fig. 10.1). This, of course, creates additional line losses, but it avoids excessive switching, and consequently reduces circuit breaker maintenance frequency.

Harmonics are another, more insidious, form of reactive power.

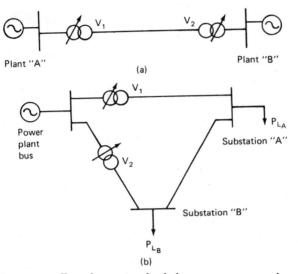

Figure 10.1 Transformer tap load changers are unevenly set ($V_1 \neq V_2$) to create a circulation current. (a) and (b) are two examples of how reactive power wheeling is accomplished.

They create substantial losses and overheat equipment. To some extent, harmonics are generated by alternators and transformers, and to a large extent by high-voltage direct-current (HVDC) transmission lines. Through filters, they are kept below a percentage that is considered acceptable.

The modern source of harmonics is solid-state electronic power equipment, especially uninterruptible power supplies and variable-speed drives. These harmonics are a problem at the distribution level. They contaminate the low- and medium-voltage grid, creating huge thermal losses. This overload is not always recognized as such.

Alternator Field Excitation Control

It is said that alternator voltage is controlled by field excitation. This is true only for a stand-alone unit in the island mode. As soon as an alternator is paralleled, its voltage is fixed by the bus to which it is connected (Fig. 10.2). Field excitation changes only the alternator's *power factor*. Only if *all* the alternators on a bus raise or lower their field excitation will the voltage be significantly affected. Both voltage and power factor are allowed to fluctuate within a small range.

Since power factor is determined by the load, the alternators' bus will have to operate at this level, and alternators will have to accept whatever the bus gives them. This is a source of instability because, if an alternator is forced to operate at high leading power factors, the

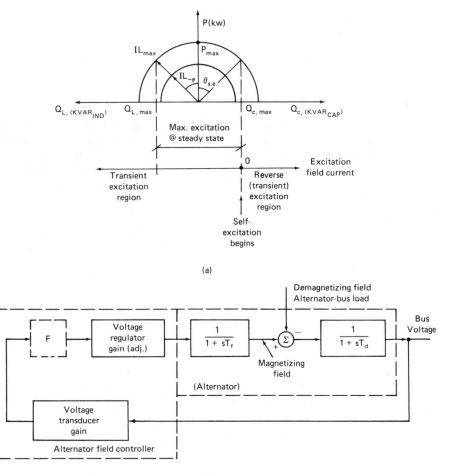

Figure 10.2 Simplified stand-alone alternator field control. (*a*) *P-Q* diagram. θ_{se}: beginning of self-excitation power factor angle; ADJ: adjustable gain parameter; θ: power factor angle; P: active load (kW), Q_c: capacitive load (KVAR, cap), Q_L: inductive load (KVAR, ind). (*b*) Block diagram.

synchronizing torque is weaker, and during a fault, loss of synchronism can occur. Leading power factors and reduced excitation can also create other stability problems, such as overvoltages and self-excitation. Reducing excitation further to lower the voltage in these cases makes the problem even worse!

Excessive lagging power factor, on the other hand, requires high field excitation currents. Field windings are already hard to cool properly, and winding damage occurs easily.

Therefore, field excitation control has two main purposes:

1. Equalize power factor among all alternators
2. Improve dynamic and transient stability

Field Control of Alternators in Parallel

The simplified block diagrams (Figs. 10.3 and 10.4) show a "demagnetizing field" as input or load. The field is represented by a complex variable that depends on a combination of the magnitude of the current I_L and the power factor of the load. When the power factor is sufficiently capacitive, self-excitation begins and the demagnetizing field is zero, regardless of the magnitude of the current. This is most undesirable since control of the voltage from the excitation field is now completely lost. Therefore, the demagnetizing field load can be approximated by

$$\text{Field (demagnetizing)} \approx I_L \times (\sin \theta_{se} - \sin \theta)$$

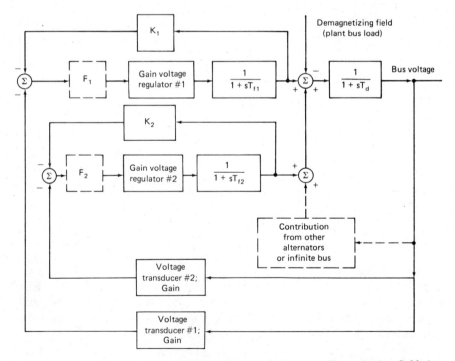

Figure 10.3 Alternators in parallel: Simplified block diagram. T_f: excitation field time constant (~ 5 s); T_d: direct reactance time constant (~ 1 s); F: first-order filter with time constant $\approx T_f$; K: load transducer gain to help even reactive load distribution.

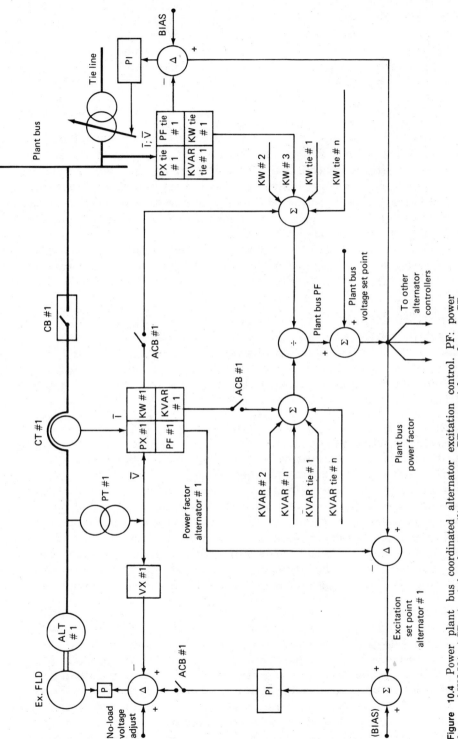

Figure 10.4 Power plant bus coordinated alternator excitation control. PF: power factor = kW/kVA; ACB: circuit breaker auxiliary contact; PT: potential transformer; CT: current transformer; PX: power system transducer; VX: voltage transducer; P: gain only; PI: gain and reset.

where θ_{se} = power factor ngle where self-excitation begins
     θ = power facto angle of load current I_L
     $|I_L|$ = load magnitude

Since most power plants consist of several alternators running in parallel, it is important to understand the dynamics of such a mode of operation. The simplified block diagram (Fig. 10.4) shows conceptually how alternators contribute to maintain the bus voltage. This scheme of things is very similar to steam boilers in parallel (Chap. 4), the differences being (1) that there may be hundreds of generators in parallel, and (2) the largest time constant is located after the controller, instead of before it.

These model representations illustrate a few stability problems that alternator voltage regulators in general, and paralleled alternators in particular, have:

1. Since the largest time constant T_f is after the controller, large controller output swings occur, with little effect on the bus voltage and questionable effect on stability in general. To mitigate this effect, a first-order filter "F," with a time constant of the same order of magnitude as the alternator field, can be installed at the input of the controller.

2. From Fig. 10.3, alternators in parallel, we observe how it is now possible to measure the individual contributions of each alternator and to introduce a stabilizing feedback signal. It is important to recognize that this feedback signal *does not stabilize the controller output significantly*, but instead helps to maintain a reactive load distribution among alternators. Nevertheless there still is a residual power factor *drift* among alternators even with this popular control scheme. Individual power factor drift among alternators occurs because of the slight discrepancies between the gains of the individual components, e.g., between the voltage transducers, load transducers, the controller settings, and also between the difference of alternators' parameters.

3. Integral action of a PID controller would further contribute to drift since each alternator now will try to "steal the show" by compounding its own contribution. Integral action for bus voltage control can only be added at the plant bus level, that is, by the addition of a master controller for the whole generating plant bus. See the power plant coordinated alternator excitation control diagram (Fig. 10.4). However, if the alternators are connected to an infinite bus, which is the case in most applications, over- or underexciting will still not affect the bus voltage significantly and instead will create stability

problems if the contribution of the infinite bus cannot be controlled via transformers fitted with load tap changers, or otherwise.

4. To eliminate power factor drift, the power factor of each generating source must be individually compared to the bus load power factor (Fig. 10.4) and adjusted accordingly. Only afterward can coordinated bus voltage control take place. To determine the bus load power factor is by no means a simple matter because it is not a direct, but a calculated measurement, since all feeder loads must be accounted for and totalized in their active and reactive components. Tie lines will count either as "generators" or "loads" depending on the direction of the flow of power. The power factor will then be

$$\text{Power factor} = \cos\left(\arctan\frac{Q}{P}\right)$$

$$= \cos\left(\arcsin\frac{Q}{|V| \times |I|}\right)$$

$$= \frac{P}{|V| \times |I|}$$

Transducer failure and measurement error must be checked to avoid controller-induced instability.

Dynamic Losses in Excitation Field Windings

Turboalternator excitation field windings are hard to cool, thus there is a lower limit to the steady-state field current that they will sustain without thermally damaging the insulation of the winding. Overexciting the field for short periods of time for stability purposes is a common practice. However, as we saw in Chap. 4 under "Energy Efficient Control," there is additional energy dissipated in the winding each time the current changes. Consequently it is possible for the winding to exceed its permissible temperature even when the average current is less than the steady-state maximum allowed.

If the disturbances the controller is trying to meet are fast enough, say, more than twice the inverse of the field time constant, then most of the control effort will dissipate as heat, without producing any useful result but only increasing the temperature in the alternator's field windings. By installing the filtering time constant "F" *before* the controller, this effect can be minimized.

Load-Voltage Time Constant

Loads also have a time constant when voltage changes. This time constant is a function of the power factor. A perfect heat load, or a power factor of 1, has no time constant, while a pure reactive load has an infinitely long time constant.

$$T_l = \frac{\tan{(\arccos{\theta})}}{\omega}$$

where $1/\omega < 0.003$ s at 60 Hz.

This means it only becomes significant at extremely low power factors that do not normally occur. During faults (short circuits) the load is so low that only the alternator winding time constants now become significant.

Transient Stability Analysis

Transient stability can be better described as *absolute* stability. Transient stability analysis (Fig. 10.5) answers the question "Will a plant remain synchronized after a fault, i.e., short circuit?" It has to do with system reliability.

Loss of synchronism can occur during a fault because, before the fault, the synchronizing torque may have been weakened to the point that alternators will fall out of step. This illustrates the importance of running alternators at stable power factors. As soon as loss of synchronism occurs, protective relays pick up a fault (i.e., overcurrent) condition and trip the alternators' breakers.

Dynamic stability can be described as *relative* stability, and is more the province of power-frequency control. Relative stability deals with fluctuations during normal operation, especially voltage and frequency, therefore it has to do with quality of service.

Synchronization

Synchronization is a critical step for successful EES operation. There are three ways of synchronizing alternators:

- Manual synchronization
- Automatic synchronization
- Self-synchronization

There are five requirements for synchronization:

- Equal voltage
- Equal frequency
- Equal phasor sequence
- Equal phasing
- System phasors must be stationary

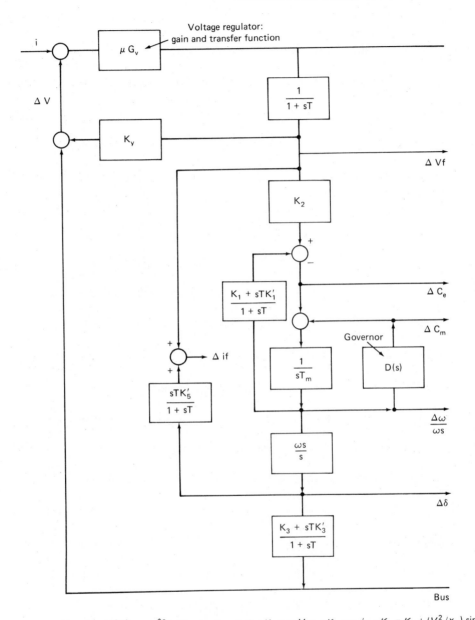

For $x_d = x_q$: $K_1 = V_0^2(\hat{q}_0 - x_p \hat{S}_0^2)$; $K_2 = V_0 /x_q \sin \theta_0$; $K_3 = x_p V_0 p_0$; $K_4 = x_p/x_q$; $K_1' = K_1 + (V_0^2/x_0) \sin \theta_0$;
$x_p = x_e x_q/(x_e + x_q)$; $x_0 = x_q^2/(x_q - x_d')(T/T_p)$; $T = (x_e + x_d'/(x_e + x_q))$; $p_0 = P_0/V_0^2$;
$\hat{q}_0 = (Q_0/V_0) + (1/x_q)$; $\hat{S}_0 = \sqrt{p_0^2 + \hat{q}_0^2}$; $\theta_0 = \arctan p_0/q_0$; $K_5' = (K_1' - K_1)K_2$; $K_3' = (K_3' - K_3)/K_4$;

P_0: kW load active; Q_0: KVAR load (reactive); x_d: Direct reactance (alternator); x_q: Quadrative reactance (alternator); x_d': Transient; x_e: Equivalent bus series reactance ($\sim$ .2); Pampers and saturation neglected: S: Laplace operator; μ: Gain; G_v: Voltage regulator transfer function; T_p: Excitation field time constant ($\sim$ 6 sec).

Figure 10.5 Generator connected to infinite bus through a series reactance. (*From G. Quazza and E. Ferrari, "Role of Power Station Control in Overall System Operation," Brown Boveri Symposium Proceedings, E. Handschin, ed., Elsevier, 1971, p. 247.*)

If these five requirements are satisfied within a certain tolerance, then synchronization will occur uneventfully.

When manual synchronization is performed, the operator adjusts the synchronizing alternator excitation for voltage and the turbine speed for frequency, and by nudging the frequency, slowly brings the phasors together until they overlap. At that point the operator manually closes the alternator-bus breaker. The alternator is now synchronized and can be loaded by opening the steam or fuel valve; since it is synchronized, the speed will not change, and only the torque will.

To avoid costly human error, a lockout "synch-check" relay checks whether all the conditions have been met within a certain range. If so, it allows closing of the circuit breaker; otherwise it will not allow closing. Equal phasor sequence is checked only the first time an alternator is paralleled. But for alternators on high-voltage systems, potential transformers are used to reduce the voltage to a manageable level, usually 110 V. This introduces a complication, since the phasing order must be maintained at the secondary of the voltage transformers, and careful attention must be devoted to the phases so as not to inadvertently invert the sequence somewhere along the control wiring.

Synchronizing out of phase is a serious mishap that causes damage to the equipment, windings, shafts and shaft couplings, the prime mover, buses, breakers, and voltage regulators. Out-of-phase synchronization can easily occur when lockout relays are not installed on *all breakers that may close on two independent phasor systems*. Sometimes it is not obvious that a bus tie breaker may be in that situation, and consequently there is no protection against such an event.

Inadvertent crossing of control wires from potential transformers by maintenance crews is also a probable cause for out-of-phase synchronization. All wires must be adequately tagged both on the wire and the termination block to avoid any confusion later on.

Automatic synchronization follows the same sequence as the manual procedure, but it is now executed by electronic controls. Automatic synchronizers for steam turbines can be finnicky, and depend heavily on the speed regulator's performance, which can be erratic and overreact at no-load conditions because the steam valves must throttle relatively little amounts of steam.

Self-synchronization is no longer used, but it has its merits. Self-synchronization does not require exact matching of frequency, voltage, or phase. Once the prime mover is brought up to synchronizing speed, the alternator bus breaker is closed and, immediately after that, the alternator excitation field breaker is closed. Self-synchronization occurs at that point. This method is no longer used because it causes additional disturbances in the alternator field excitation that can create temporarily unstable power factor conditions.

Self-synchronization consists of connecting the alternator to the bus as an asynchronous, or induction, generator. It is an attractive alternative for small, rugged, unsophisticated power plants, and where skilled labor is unavailable. Self-synchronization eliminates the risk of out-of-phase synchronization and consequent serious damage to mechanical and electrical equipment alike.

Area Synchronization

When it is desirable or necessary to synchronize two different areas, the same considerations as before apply. Synchronizing tens or hundreds of units together poses above all an organizational problem to get the frequencies of both systems to move in the right direction.

Another problem with area synchronization is the "stiffness" of the tie. In order for the areas to *stay* synchronized, enough synchronizing power (torque) must be allowed to circulate between them. If the tie between two areas does not supply enough synchronizing torque, there will be loss of synchronism, and protective relays will trip the interconnection. If more than one tie line is required to provide enough synchronizing torque, then a simultaneous closing procedure for tie lines must be put in place. This may take hours, or days, to set up, hence the reluctance of load dispatch operators to go into island mode. The following equation[1] estimates the lower size limit that a tie line must have in order to be successful. It is based on statistical considerations and assumes reasonably adjusted governors:

$$P_{\text{tie}} = 0.43 \sqrt{\frac{P_1 P_2}{P_1 + P_2}}$$

where P_1 and P_2 are the capacities of the systems to be joined.[1]

HVDC Synchronization

The solution to many of these problems is the HVDC transmission line. DC transmission is frequency-independent. In other words, it does not provide or require synchronizing torque, but still controls the flow of energy. An HVDC line must allow *some* energy to flow; it cannot stay unloaded below a certain design limit, approximately 40 percent of maximum capacity. HVDC transmission has the following advantages:

- No fault currents are transmitted (instantaneous fault clearing)
- No distance limitations for either overhead or underground lines
- No synchronicity requirements (frequency, phase, or voltage)
- Improves stability by controlling power flow

HVDC transmission has the following disadvantages:

- Generates harmonics
- Cannot be networked
- Is expensive

HVDC lines are highly inductive ac loads. They also require comparable generation at the receiving end in order to function; i.e., they cannot transmit 100 percent of the user's load, since they need to be "excited" at the receiving end.

Dynamics of HVDC Transmission[2]

An HVDC system can be best described as a hybrid between a transmission line and a power plant. It can both transport and control the flow of power. HVDC transmission plays an important role in improving system stability. This is because there is no synchronicity requirement and it provides positive oscillation damping.

Thus, HVDC transmission is exempt from the synchronicity requirement to which all alternators, ac power lines, and transformers are subject. Synchronicity becomes a problem when the interconnected system becomes very large. Disturbances, dead bands in governors, and weak tie lines create unstable modes that result eventually in the separation of parts of the system and possible blackouts.

HVDC's positive oscillation damping cancels unstable modes created by weaker parts of the system. This is possible because an HVDC link can control the flow of power much faster than any power plant or phase transformer—i.e., instantaneously.

In addition, HVDC transmission is very flexible. It can interconnect systems with different frequencies (50 and 60 Hz, for example) and it can limit power transfer to keep systems separated in a back-to-back tie, with no transmission line in between.

The following are possible uses of HVDC:

- Interconnect two large systems (England and France)
- Transport energy over long distances (Manitoba)
- Connect systems having different frequencies (Japan)
- Reduce losses over long distances (Pacific intertie)
- Isolate—i.e., limit energy transfer—between systems (Texas)
- Supply energy to congested urban areas in large amounts

Typical HVDC Arrangements

Monopolar systems (Fig. 10.6) with ground return are normally not allowed, unless they run over seawater, since they may produce disturbances in railway systems and electrolytic corrosion in underground structures.

Bipolar systems (Fig. 10.7) are normally used. In them, the rectifiers/inverters are connected in series and the ac transformers are connected in parallel. There are two conductors: positive and negative. The middle point of the dc system is grounded, but normally there is no current circulation. Current circulates only through the conductors, unless there is a fault, in which case circulation is allowed for short periods of 30 min or so.

Typical design data is as follows:

Voltage	+ 100 kV to + 500 kV
Power	Up to 4000 MW
Current	1000 to 2000 A dc
Distance	300 to 1000 mi

For 1000-mi-long overhead transmission lines, the voltage is 300 to 500 kV. For underground cable lines 20 to 100 miles long, it is 100 kV.

Advantages

HVDC transmission limits short-circuit currents, since the valves, or thyristors, can be shut down instantaneously. The first half-cycle is

Figure 10.6   Monopolar HVDC system.

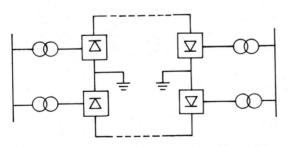

Figure 10.7   Bipolar HVDC system (normally used).

anot limited, but everything after that is. (In HVAC the best circuit breaker available operates after three cycles, and the subtransient current is not limited, only the transient period.)

DC transmission lines can be represented by a resistance. There are no charging current, voltage increase, and length limitations, as there are in ac.

The voltage-drop regulation for dc is

$$\Delta V = IR$$

and for ac is

$$\Delta V = IR + j\omega LI$$

That is, the voltage drop is much smaller in dc, because there is no inductive load.

In HVDC there is no stability problem.

The dc line is able to carry the same amount of power as an ac line but only two conductors are required instead of three, therefore it is less expensive than an ac line.

Disadvantages

HVDC transmission generates harmonics in the ac system (the third, fifth, eleventh, and thirteenth, etc.). There are no HVDC circuit breakers available, therefore networking is not possible. Rectifiers and inverters are expensive. Rectifiers and inverters use reactive power, therefore they need synchronous condensers or capacitors.

General Description of an HVDC System

In a representative HVDC system (Fig. 10.8), circuit breakers (CB) are used to interrupt the system or to remove a fault, while the disconnect switches (DS) are intended to ensure safety by physically separating and showing their separation, to service the circuit breakers.

The lightning arrester (LA) protects the transformer against lightning strokes in the ac line. The ac filter is tuned to eliminate—i.e., short-circuit—higher frequencies or harmonics generated by the dc system.

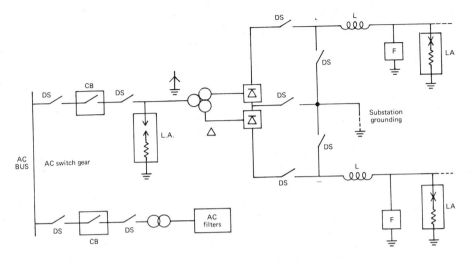

Figure 10.8 Typical HVDC station. DS: disconnect switch; CB: circuit breaker; LA: lightning arrester; F: harmonic filter.

The converter transformer can be a single unit with a third winding delta instead of two separate transformers, which are more expensive. The D-Y (delta-wye) connection is used to produce a 30° phase shift.

The delta winding in the two-transformer arrangement is used to reduce the zero-sequence impedance. Sometimes it is used to supply power to the ac harmonic filters.

The converter is a solid-state device using several thousand thyristors. (Older systems used mercury-arc devices.) The converter can operate as a rectifier or as an inverter. It always uses the three-phase bridge connection (Fig. 10.9). It consists of six thyristor-based valves (or mercury valves). Several other connections have been proposed, but this is the only one actually used. The anode reactance "L" resembles that of a transformer and is approximately 0.1 H (it used to be around 1 H). It is used to smooth and maintain the dc constant.

The dc switchgear is used to short-circuit the converter or remove the converter from the circuit. It can be used only when the system is not in operation.

The dc filters are used to eliminate the harmonics of the dc side (sixth, eighth, tenth, etc.)—i.e., even harmonics. One or two tuned filters and one high-pass filter are used to short-circuit them. These harmonics must be eliminated because they induce interference in telephone lines.

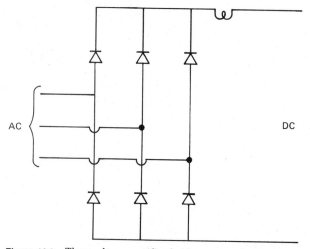

Figure 10.9   Three-phase rectifier bridge connection.

The lightning arrestor is used to reduce or eliminate the overvoltage produced by lightning on the dc line. In addition, the HVDC transmission line has a disconnect switch.

The substation grounding is a ground net-mat of iron rods, welded at every node. Everything is connected to this ground mat.

For ground transmission there is a ground electrode separated 2 to 3 mi from the substation, in a remote, inaccessible area. It must be fenced off to keep out unauthorized people and protect them from electrical shock. If seawater is nearby, the electrode is installed there. This electrode grounds the middle point of the dc system and is designed to carry the full 2000- to 3000-A current. It is connected to the middle point of the dc system via a single isolated transmission line, with insulation that can withstand up to 50 kV.

Converter Operation

The heart of the HVDC system is the six thyristor valves. A valve is basically a switch that can be turned on by an electric signal applied to the gate electrode but cannot be turned off after that. It turns off only when the current goes through zero and the polarity changes, since the current can go through the valve in only one direction, i.e., from anode to cathode.

Each valve consists of several hundred thyristors connected in series, together with a compensating network (Fig. 10.10). Each gate is supplied a pulse through an isolating transformer (or optical fiber in the latest technology).

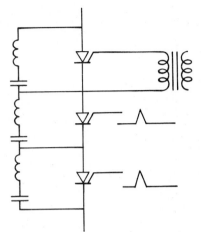

Figure 10.10 HVDC thyristor valve arrangement.

Inverter Operation

Delaying the firing of the thyristor valves keeps the dc voltage negative so that the power will flow from dc to ac. The inversion occurs by delaying the firing 150 to 170°. This is possible only if the outside ac power source generates and maintains the sine wave. A 10 to 15° margin must be maintained below 180° phase shift because of valve operational requirements, which in turn generate additional losses and decrease their controllable range.

HVDC System Control

The objective of HVDC line operation is to maintain the power factor as high or as close to stability as possible. Power flow can be controlled by the constant-current or the constant-voltage method. To keep the current constant, the rectifier firing angle is changed; to keep the voltage constant, the inverter firing angle is changed. Quasi-constant-voltage operation gives the best results. There is an additional current while voltage is changed because of the capacitance of the dc line.

Reference

1. C. Concordia, "Effect of Prime Mover Speed Control Characteristics on Electric Power System Performance," *IEEE Transactions on Power Systems*, vol. PAS-88, no. 5, May 1969, pp. 752–756.
2. Notes from New York Polytechnic HVDC seminar, 1985.

Typical Logic Diagrams and Instrumentation and Control Diagrams for Boiler Plants in General

GENERAL NOTES:

1. INSTRUMENTATION & CONTROL DIAGRAMS ARE TO BE USED IN CONJUNCTION WITH THE INSTRUMENTATION AND CONTROL EQUIPMENT LIST

2. REPRESENTATION OF ANALOG AND DIGITAL CONTROLS IS FUNCTIONALLY ORIENTED. ACTUAL HARDWARE IMPLEMENTATION MAY DIFFER FROM THAT SHOWN ON ICD'S.

3. SCOPE OF SUPPLY IS NOT TO BE DETERMINED FROM ICD'S. REFER TO INSTRUMENT & CONTROL EQUIPMENT LIST AND SPECIFICATION DOCUMENTS.

4. △ =SIGNIFIES TRAIN NUMBER IN INDENTIFICATION CODE
   SPECIFICALLY.

   1 FOR TRAIN 1
   2 FOR TRAIN 2
   3 FOR TRAIN 3

5. ALL HAND SWITCHES ARE MOMENTARY CONTACT PUSHBUTTONS (NORMALLY OPEN, CLOSED WHEN DEPRESSED) UNLESS OTHERWISE STATED ON THE LOGIC DIAGRAMS.

6. UNLESS OTHERWISE NOTED, ALL MOTOR OPERATED VALVES WILL GO FULL TRAVEL UNTIL STOPPED BY LIMIT SWITCH ACTION IN OPEN POSITION & BY TORQUE SWITCH IN CLOSED POSITION. VALVE TRAVEL IS STOPPED IN INTERMEDIATE POSITION BY MOTOR OVERLOAD OR HIGH TORQUE. BOTH RED & GREEN LIGHTS WILL BE ENERGIZED (ON) FOR A VALVE IN INTERMEDIATE POSITION.

7. THE LIMIT SWITCH & TORQUE SWITCH INTERLOCKS ARE HARD WIRED AT THE MOTOR STARTER.

8. UNLESS OTHERWISE NOTED, CONTACTS TO INDICATE THE CONTROL VALVE OPEN, CLOSE OR MODULATING, ARE LIMIT SWITCHES ON THE CONTROL VALVE WITH GREEN LIGHT "ON" BY LIMIT SWITCH, WHICH IS CLOSED IN ALL POSITIONS, EXCEPT WHEN THE CONTROL VALVE IS FULLY OPEN THE RED LIGHT IS "ON" BY LIMIT SWITCH WHICH IS CLOSED IN ALL POSITIONS EXCEPT WHEN THE VALVE IS FULLY CLOSED. BOTH GREEN & RED LIGHTS WILL BE "ON" WHEN THE CONTROL VALVE IS MODULATING.

SOLID WASTE RESOURCE RECOVERY FACILITY

GENERAL NOTES
AND REFERENCES

Gibbs & Hill, Inc.
ENGINEERS, DESIGNERS, CONSTRUCTORS
New York

SCALE: NONE

B1C-0200

SH. NO. 01

JOB NO.

FILM NO.	ISSUE NO.	DATE	DWN.	CHKD.									
O	10/14/88	RLR	SMM		AGD/SNC	-	SMM	-	HCS	CONFIGURATION		
A	3/8/88	RLR	AJS		AGD/HS	-	SMM	-	FLG	CLIENTS REVIEW		
MICROFILM		DATE	DWN.	CHKD.	ARCH-STR	MECH-ELEC	I&C	SHLD/CONST P.M.		ISSUED FOR		
						APPROVALS						

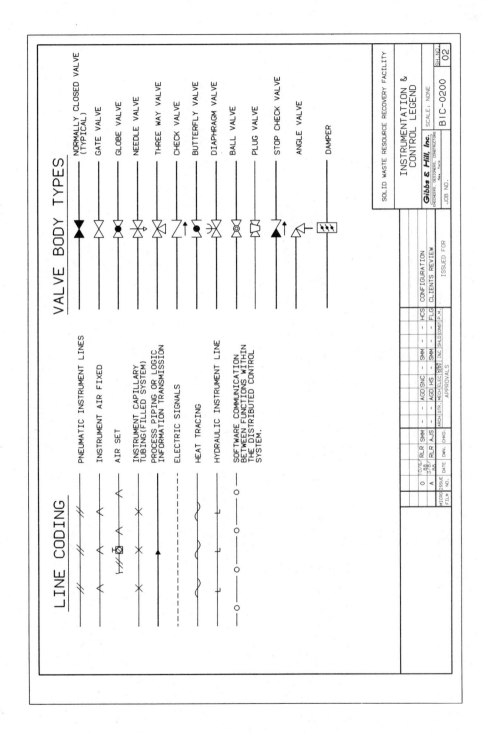

LINE CODING

—//—//—	PNEUMATIC INSTRUMENT LINES
—∧—∧—∧—	INSTRUMENT AIR FIXED
—//—⊠—∧—	AIR SET
—×—×—×—	INSTRUMENT CAPILLARY TUBING (FILLED SYSTEM)
—▲—	PROCESS PIPING OR LOGIC INFORMATION TRANSMISSION
— — — — —	ELECTRIC SIGNALS
—ʅ—ʅ—ʅ—	HEAT TRACING
—~—~—	HYDRAULIC INSTRUMENT LINE
—o—o—o—	SOFTWARE COMMUNICATION BETWEEN FUNCTIONS WITHIN THE DISTRIBUTED CONTROL SYSTEM.

VALVE BODY TYPES

▶◀	NORMALLY CLOSED VALVE (TYPICAL)
▷◁	GATE VALVE
▷●◁	GLOBE VALVE
▷⧸	NEEDLE VALVE
▷◁⧸	THREE WAY VALVE
◁↑	CHECK VALVE
▷●◁	BUTTERFLY VALVE
▷⧸◁	DIAPHRAGM VALVE
▷⊗◁	BALL VALVE
▷◁	PLUG VALVE
▶◁↑	STOP CHECK VALVE
◁↑	ANGLE VALVE
▤	DAMPER

SOLID WASTE RESOURCE RECOVERY FACILITY

INSTRUMENTATION & CONTROL LEGEND

Gibbs & Hill, Inc.
ENGINEERS, DESIGNERS, CONSTRUCTORS
New York

SCALE: NONE

SH. NO. 02

B1C-0200

JOB NO.

FILM NO.	ISSUE NO.	DATE	DWN.	CHKD.							
O	10/6/88	RLR	SMM		ARCH	STR.	MECH	ELEC	INST	CONFIGURATION	HCS
A	3/8/88	RLR	AJS		AGD	SNC	SMM	SMM		CLIENTS REVIEW	FLG
					AGD	HS	SMM	DONG.	P.M.	ISSUED FOR	

MICRO FILM

APPROVALS

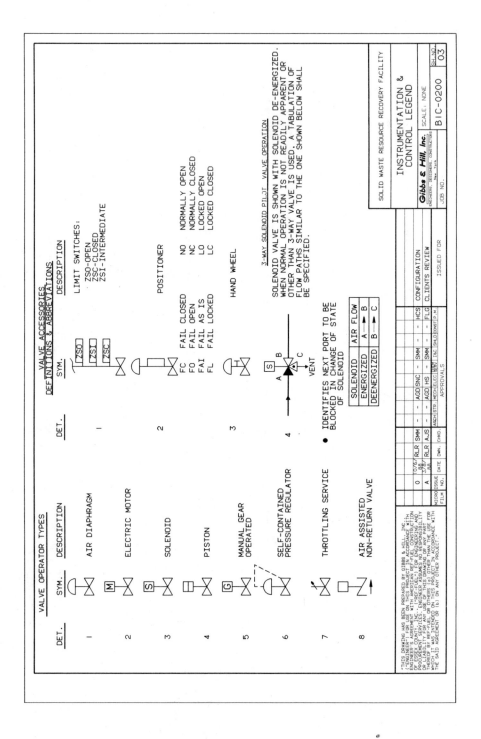

VALVE OPERATOR TYPES

DET.	SYM.	DESCRIPTION
1		AIR DIAPHRAGM
2	M	ELECTRIC MOTOR
3	S	SOLENOID
4		PISTON
5	G	MANUAL GEAR OPERATED
6		SELF-CONTAINED PRESSURE REGULATOR
7		THROTTLING SERVICE
8		AIR ASSISTED NON-RETURN VALVE

VALVE ACCESSORIES DEFINITIONS & ABBREVIATIONS

DET.	SYM.	DESCRIPTION
1	ZSO ZSI ZSC	LIMIT SWITCHES: ZSO-OPEN ZSC-CLOSED ZSI-INTERMEDIATE
2		POSITIONER
	FC FO FAI FL	FAIL CLOSED FAIL OPEN FAIL AS IS FAIL LOCKED
	NO NC LO LC	NORMALLY OPEN NORMALLY CLOSED LOCKED OPEN LOCKED CLOSED
3		HAND WHEEL

3-WAY SOLENOID PILOT VALVE OPERATION

4	

SOLENOID VALVE IS SHOWN WITH SOLENOID DE-ENERGIZED. WHEN NORMAL OPERATION IS NOT READILY APPARENT OR OTHER THAN 3-WAY VALVE IS USED, A TABULATION OF FLOW PATHS SIMILAR TO THE ONE SHOWN BELOW SHALL BE SPECIFIED.

• IDENTIFIES NEXT PORT TO BE BLOCKED IN CHANGE OF STATE OF SOLENOID

SOLENOID	AIR FLOW
ENERGIZED	A → B
DEENERGIZED	B → C

SOLID WASTE RESOURCE RECOVERY FACILITY

Gibbs & Hill, Inc.
ENGINEERS, DESIGNERS, CONSTRUCTORS
New York

INSTRUMENTATION & CONTROL LEGEND

SCALE: NONE

JOB NO.

BIC-0200

SH. NO. 03

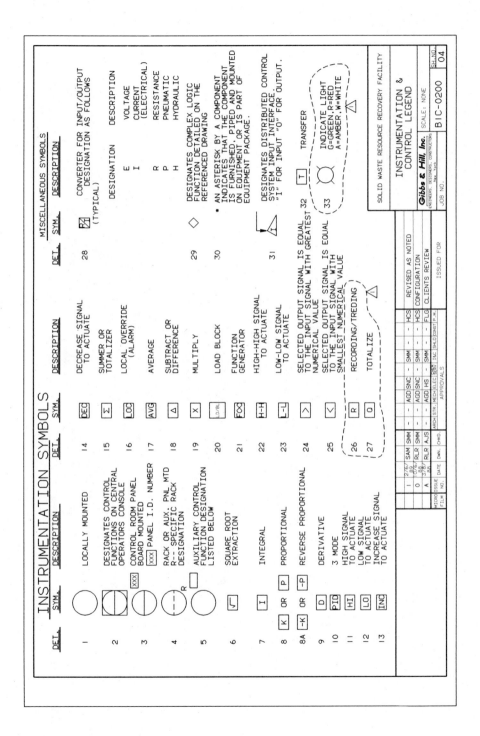

INSTRUMENTATION SYMBOLS

DET.	SYM.	DESCRIPTION
1	(circle)	LOCALLY MOUNTED
2	(circle w/ line)	DESIGNATES CONTROL FUNCTIONS ON CENTRAL OPERATORS CONSOLE
3	(circle, xxx)	CONTROL ROOM PANEL BOARD MOUNTED [xxx] PANEL I.D. NUMBER
4	(dashed circle, R)	RACK OR AUX. PNL MTD R-- SPECIFIC RACK DESIGNATION
5	(rectangle)	AUXILIARY CONTROL FUNCTION DESIGNATION LISTED BELOW
6	√	SQUARE ROOT EXTRACTION
7	I	INTEGRAL
8	K OR P	PROPORTIONAL
8A	-K OR -P	REVERSE PROPORTIONAL
9	D	DERIVATIVE
10	PID	3 MODE
11	HI	HIGH SIGNAL TO ACTUATE
12	LO	LOW SIGNAL TO ACTUATE
13	INC	INCREASE SIGNAL TO ACTUATE
14	DEC	DECREASE SIGNAL TO ACTUATE
15	Σ	SUMMER OR TOTALIZER
16	LOC	LOCAL OVERRIDE (ALARM)
17	AVG	AVERAGE
18	Δ	SUBTRACT OR DIFFERENCE
19	X	MULTIPLY
20	LD/BL	LOAD BLOCK
21	f(x)	FUNCTION GENERATOR
22	H-H	HIGH-HIGH SIGNAL TO ACTUATE
23	L-L	LOW-LOW SIGNAL TO ACTUATE
24	>	SELECTED OUTPUT SIGNAL IS EQUAL TO THE INPUT SIGNAL WITH GREATEST NUMERICAL VALUE
25	<	SELECTED OUTPUT SIGNAL IS EQUAL TO THE INPUT SIGNAL WITH SMALLEST NUMERICAL VALUE
26	R	RECORDING/TREDING
27	Q	TOTALIZE

MISCELLANEOUS SYMBOLS

DET.	SYM.	DESCRIPTION
28		CONVERTER FOR INPUT/OUTPUT DESIGNATION AS FOLLOWS (TYPICAL)

DESIGNATION	DESCRIPTION
E	VOLTAGE
I	CURRENT (ELECTRICAL)
R	RESISTANCE
P	PNEUMATIC
H	HYDRAULIC

DET.	SYM.	DESCRIPTION
29	◇	DESIGNATES COMPLEX LOGIC FUNCTION DETAILED ON THE REFERENCED DRAWING
30	*	AN ASTERISK BY A COMPONENT INDICATES THAT THE COMPONENT IS FURNISHED, PIPED AND MOUNTED ON EQUIPMENT OR IS PART OF EQUIPMENT PACKAGE.
31	(triangle)	DESIGNATES DISTRIBUTED CONTROL SYSTEM INPUT INTERFACE. "I" FOR INPUT "O" FOR OUTPUT.
32	T	TRANSFER
33	(circle)	INDICATE LIGHT G=GREEN, R=RED A=AMBER, W=WHITE

SOLID WASTE RESOURCE RECOVERY FACILITY

Gibbs & Hill, Inc.
ENGINEERS, DESIGNERS, CONSTRUCTORS

INSTRUMENTATION & CONTROL LEGEND

SCALE: NONE

JOB NO.

B I C-0200

SH. NO. 04

MEANING OF IDENTIFICATION LETTERS

	FIRST LETTER MEASURED OR INITIATED VARIABLE	SECOND LETTER READOUT OR PASSIVE FUNCTION	THIRD/FOURTH LETTER READOUT OR PASSIVE FUNCTION
A	ANALYSIS	ALARM	
B	BURNER/FLAME	BI-STABLE	
C	CONDUCTIVITY	CONTROL	
D	DENSITY OR SPECIFIC GRAVITY	DIGITAL COMPUTER OR DIFFERENTIAL	
E	VOLTAGE OR EMF	PRIMARY ELEMENT	ELECTRICAL
F	FLOW	FACTOR	
G	GAGING	GLASS	
H	HAND		HIGH
I	CURRENT	INDICATOR	INDICATOR
J	POWER		
K	TIME OR TIME SCHEDULE	CONTROL STATION	CONTROL STATION
L	LEVEL	LIGHT	LOW
M	MOISTURE OR HUMIDITY		
N	MOTOR RUNNING		
O		ORIFICE (RESTRICTION)	
P	PRESSURE OR VACUUM	POINT	
Q	ENERGY	POWER SUPPLY OR TOTALIZER	
R	RATIO, REVERSING	RECORD OR PRINT OR REDUCING	
S	SPEED, FREQUENCY OR SOLENOID	SWITCH, SUPPLY	SWITCH
T	TEMPERATURE	TRANSMITTER	TRANSMITTER
U	MULTIVARIABLE	MULTIFUNCTION	
V	VISCOSITY OR VIBRATION	VALVE DAMPER OR LOUVER	
W	UNCLASSIFIED	WELL,WATT	
X	TRIP	UNCLASSIFIED	
Y	STATUS	SIGNAL MODIF. RELAY OR COMPUTE	
Z	POSITION		

SOLID WASTE RESOURCE RECOVERY FACILITY

INSTRUMENTATION &
CONTROL SYMBOLS

Gibbs & Hill, Inc.
ENGINEERS, DESIGNERS, CONSTRUCTORS
New York

JOB NO.

B1C-0200

SCALE: NONE

SH NO. 05

				SAM	SMM	-	-	AGD	SNC	-	SMM	-	-	FLG	REVISED AS NOTED
O		RLR	SMM	-	-	AGD	SNC	-	SMM	-	-	FLG	CONFIGURATION
A		RLR	AJS	-	-	AGD	HS	-	SMM	-	-	FLG	CLIENTS REVIEW

ISSUE NO. | DATE | DWN. | CHKD. | ARCH | STR. | MECH | ELEC | SSM | I&C | SWLD | CONST | P.M. | ISSUED FOR

APPROVALS

LOGIC SYMBOLS

DET.	SYM.	DESCRIPTION
1	A B C → AND → X	LOGIC OUTPUT X EXISTS IF AND ONLY IF ALL LOGIC INPUTS A,B,C, EXIST.
2	A B C → OR → X	LOGIC OUTPUT X EXISTS IF AND ONLY IF ONE OR MORE LOGIC INPUTS A,B,C, EXIST.
3	A → NOT → X	LOGIC OUTPUT X EXISTS IF AND ONLY IF LOGIC INPUT A DOES NOT EXIST.
4	A B → S R (+) → X Y MAINTAINED MEMORY	S REPRESENTS "SET MEMORY" R REPRESENTS "RESET MEMORY" LOGIC OUTPUT X EXISTS AS SOON AS LOGIC INPUT A EXISTS. X CONTINUES TO EXIST REGARDLESS OF THE SUBSEQUENT STATE OF A UNTIL THE MEMORY IS RESET (TERMINATED) BY LOGIC INPUT B EXISTING LOGIC OUTPUT Y IF USED EXISTS WHEN X DOES NOT EXIST AND Y DOES NOT EXIST WHEN X EXISTS. NOTE: • OUTPUT Y IS NOT SHOWN IF IT IS NOT USED. (+) INDICATES OVERRIDING FUNCTION

DET.	SYM.	DESCRIPTION
5	A → TD (---) (t) → X ADJUSTABLE TIME DELAY	(---) SHALL BE CODE LETTERS AS DEFINED BELOW FOR MODE OF TIMER OPERATION. (t) IS THE NORMAL DELAY TIME WITH APPROPRIATE UNITS (SEC.,MIN.,HR.).
6	A → TD PU (t) → X ADJUSTABLE TIME DELAY PICK-UP	PU - PICK-UP THE CONTINUOUS EXISTANCE OF LOGIC INPUT A FOR A TIME (t) CAUSES X TO EXIST WHEN (t) EXPIRES X TERMINATES WHEN A TERMINATES.
7	A → TD DO (t) → X ADJUSTABLE TIME DELAY DROP-OUT	DO - DROP-OUT THE INITIATION OF A CAUSES X TO EXIST IMMEDIATELY. X TERMINATES WHEN (t) EXPIRES AFTER A TERMINATES. NOTE: OTHER MODES OF TIMER OPERATION SHALL BE HANDLED BY APPROPRIATE EXPLANATORY NOTES.
8	A → SS (+) → X	SINGLE SHOT THE INITIATION OF A CAUSES X TO EXIST FOR A TIME (t) ONLY. X TERMINATES WHEN (t) EXPIRES OR A TERMINATES WHICHEVER TAKES PLACE FIRST.

SOLID WASTE RESOURCE RECOVERY FACILITY

LOGIC SYMBOLS

Gibbs & Hill, Inc.
ENGINEERS, DESIGNERS, CONSTRUCTORS
New York

SCALE: NONE   B1C-0200   SH. NO. 06

JOB NO.

ISSUED FOR

CONFIGURATION   CLIENTS REVIEW

APPROVALS

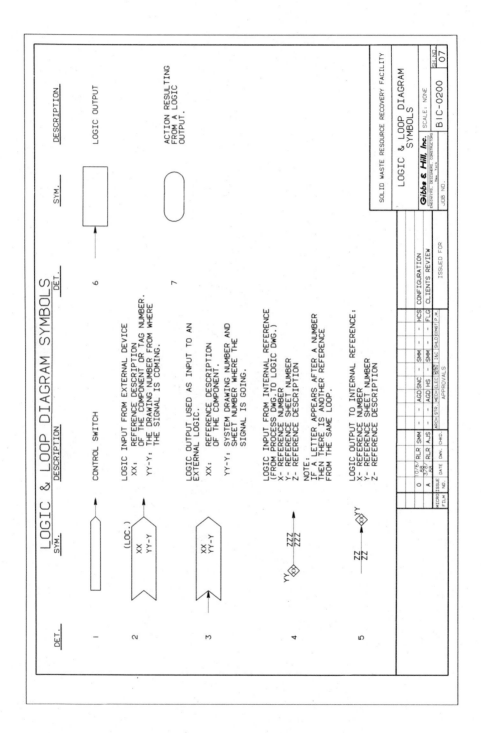

LOGIC & LOOP DIAGRAM SYMBOLS

DET.	SYM.	DESCRIPTION
1		CONTROL SWITCH
2	(LOC.) XX YY-Y	LOGIC INPUT FROM EXTERNAL DEVICE XX: REFERENCE DESCRIPTION OF THE COMPONENT OR TAG NUMBER. YY-Y: THE DRAWING NUMBER FROM WHERE THE SIGNAL IS COMING.
3	XX YY-Y	LOGIC OUTPUT USED AS INPUT TO AN EXTERNAL LOGIC. XX: REFERENCE DESCRIPTION OF THE COMPONENT. YY-Y: SYSTEM DRAWING NUMBER AND SHEET NUMBER WHERE THE SIGNAL IS GOING.
4	YY ZZZ ZZZ	LOGIC INPUT FROM INTERNAL REFERENCE (FROM PROCESS DWG. TO LOGIC DWG.) X- REFERENCE NUMBER Y- REFERENCE SHEET NUMBER Z- REFERENCE DESCRIPTION NOTE: IF A LETTER APPEARS AFTER A NUMBER THEN THERE IS ANOTHER REFERENCE FROM THE SAME LOOP.
5	ZZ ZZ YY	LOGIC OUTPUT TO INTERNAL REFERENCE: X- REFERENCE NUMBER Y- REFERENCE SHEET NUMBER Z- REFERENCE DESCRIPTION

DET.	SYM.	DESCRIPTION
6		LOGIC OUTPUT
7		ACTION RESULTING FROM A LOGIC OUTPUT.

SOLID WASTE RESOURCE RECOVERY FACILITY

LOGIC & LOOP DIAGRAM SYMBOLS

Gibbs & Hill, Inc.
ENGINEERS, DESIGNERS, CONSTRUCTORS
New York

		JOB NO.		SCALE: NONE		SH.NO.
					B I C-0200	07

					ARCH.	STR.	MECH.	ELEC.	I&C	SH&D	CONST.	P.M.					
				APPROVALS													

APPROVALS

						ARCH.	STR.	MECH.	ELEC.	I&C	SH&D	CONST.	P.M.	HCS	CONFIGURATION
0	10/6/88	RLR	SMM	-	AGD/SNC	-	SMM	-			FLG	CLIENTS REVIEW
A	3/8/88	RLR	AJS	-	AGD/HS	-	SMM	-				ISSUED FOR

| MICRO | ISSUE | DATE | DWN. | CHKD. | | | | | | | | |
| FILM | NO. | | | | | | | | | | | |

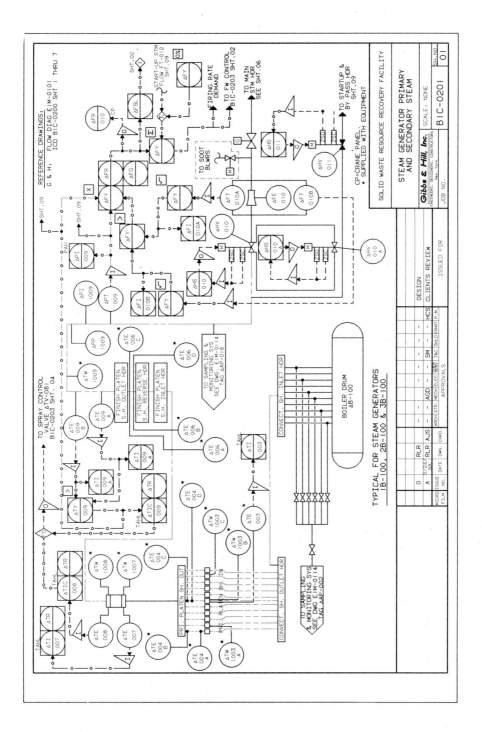

TYPICAL FOR STEAM GENERATORS
1B-100, 2B-100 & 3B-100

REFERENCE DRAWINGS:
G & H: FLOW DIAG E1M-0101
ICD B1C-0200 SHT 1 THRU 7

SOLID WASTE RESOURCE RECOVERY FACILITY

Gibbs & Hill, Inc.
ENGINEERS, DESIGNERS, CONSTRUCTORS
New York

STEAM GENERATOR PRIMARY
AND SECONDARY STEAM

| SCALE: NONE | | B1C-0201 | SH. NO. 01 |

JOB NO.

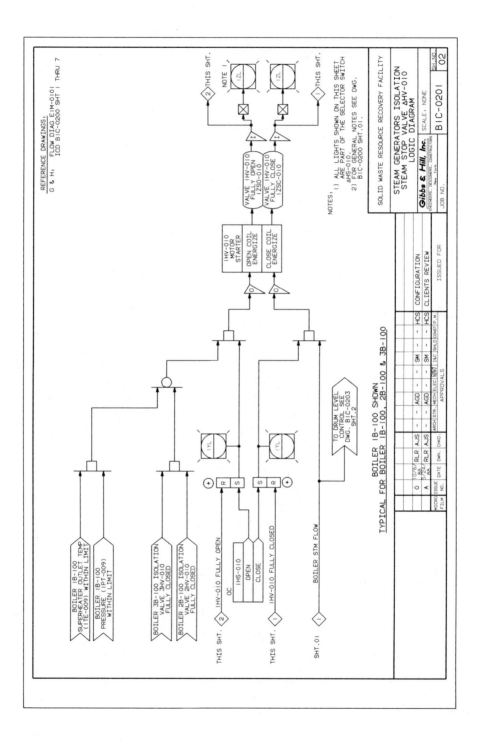

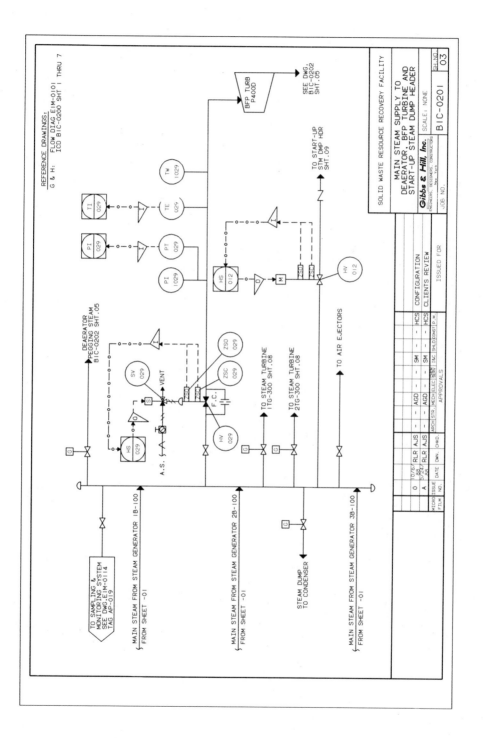

REFERENCE DRAWINGS:
G & H: FLOW DIAG. EIM-0101
ICD BIC-0200 SHT. I THRU 7

BFP TURB
P400D

SEE DWG.
BIC-0202
SHT.05

TO START-UP
STM DMP HDR
SHT.09

TW
1029

TI
029

TE
029

PI
029

PT
029

HS
012

PI
1029

ZSO
ZSD

HV
012

DEAERATOR
PEGGING STEAM
BIC-0202 SHT.05

SV
029

VENT

ZSO
029

ZSC
029

F.C.

HS
029

A.S.

HV
029

TO STEAM TURBINE
1TG-300 SHT.08

TO STEAM TURBINE
2TG-300 SHT.08

TO AIR EJECTORS

TO SAMPLING &
MONITORING SYSTEM
SEE DWG. EIM-0114
TAG AP-019

MAIN STEAM FROM STEAM GENERATOR 1B-100
FROM SHEET -01

MAIN STEAM FROM STEAM GENERATOR 2B-100
FROM SHEET -01

STEAM DUMP
TO CONDENSER

MAIN STEAM FROM STEAM GENERATOR 3B-100
FROM SHEET -01

SOLID WASTE RESOURCE RECOVERY FACILITY

MAIN STEAM SUPPLY TO
DEAERATOR, BFP TURBINE AND
START-UP STEAM DUMP HEADER

Gibbs & Hill, Inc.
ENGINEERS, DESIGNERS, CONSTRUCTORS
New York

SCALE: NONE

JOB NO.

BIC-0201

SH. NO.
03

| | | | HCS | CONFIGURATION |
| | | | HCS | CLIENTS REVIEW |

ISSUED FOR

| | | | SM | |
| | | | SM | |

0	10/6/88	RLR	AJS	-	-	AGD	ARCH. STR.	MECH/ELEC.	BEM	I&C	SH.D./CONST.	P.M.
A	5/24/88	RLR	AJS	-	-	AGD						
MICRO/ISSUE	DATE	DWN.	CHKD.			APPROVALS						
FILM	NO.											

167

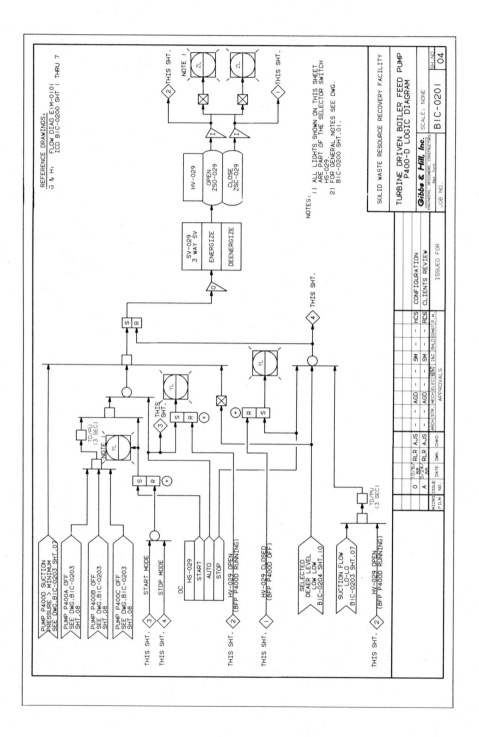

168

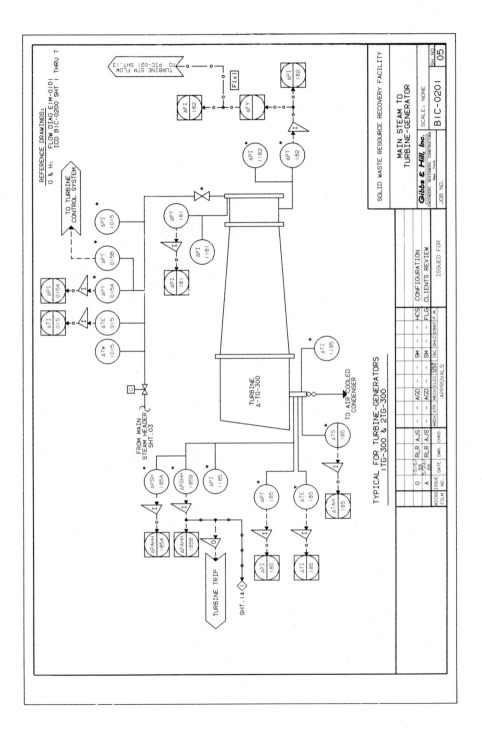

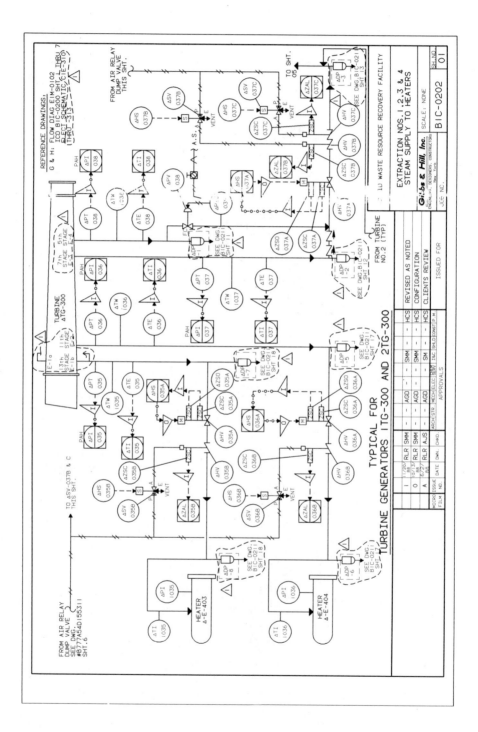

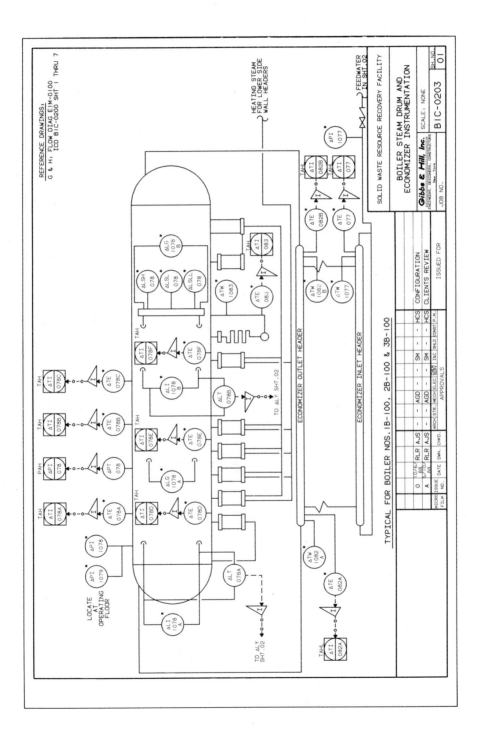

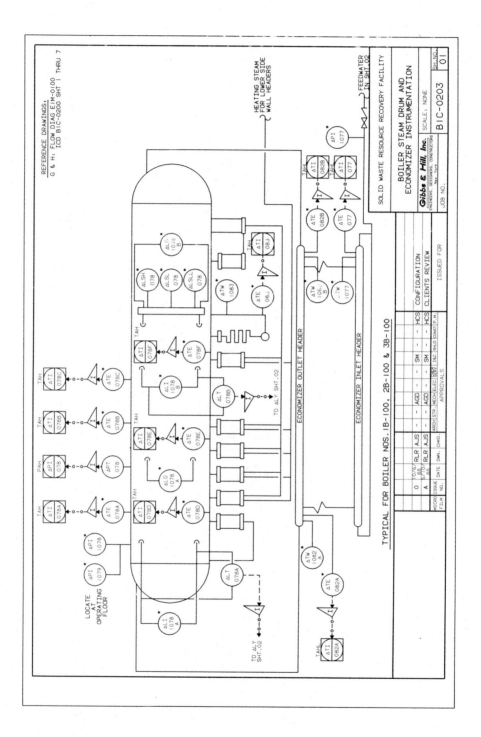

TYPICAL FOR BOILER NOS. 1B-100, 2B-100 & 3B-100

REFERENCE DRAWINGS:
G & H: FLOW DIAG EIM-0100
ICD BIC-0200 SHT 1 THRU 7

HEATING STEAM
FOR LOWER SIDE
WALL HEADERS

FEEDWATER
IN SHT.02

LOCATE
AT
OPERATING
FLOOR

ECONOMIZER OUTLET HEADER

ECONOMIZER INLET HEADER

SOLID WASTE RESOURCE RECOVERY FACILITY

BOILER STEAM DRUM AND
ECONOMIZER INSTRUMENTATION

Gibbs & Hill, Inc.
ENGINEERS, DESIGNERS, CONSTRUCTORS
New York

SCALE: NONE

BIC-0203

SH.NO. 01

JOB NO.

CONFIGURATION

CLIENTS REVIEW

ISSUED FOR

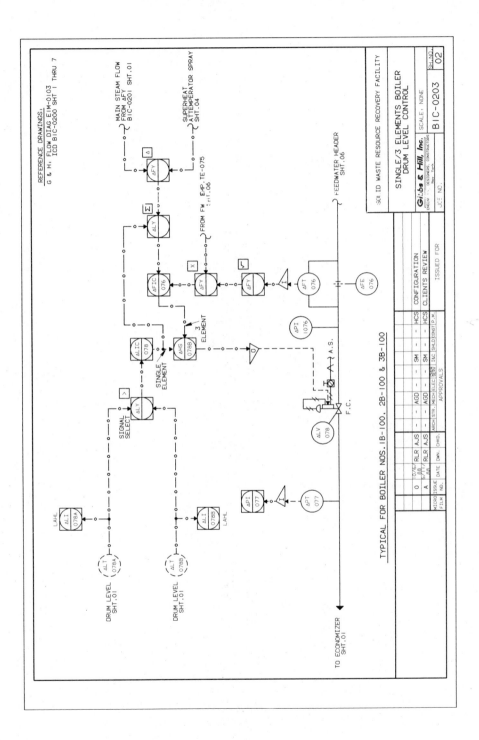

REFERENCE DRAWINGS:
G & H: FLOW DIAG E1M-0103
ICD B1C-0200 SHT 1 THRU 7

MAIN STEAM FLOW
FROM ΔFY
B1C-0201 SHT.01

SUPERHEAT
ATTEMPERATOR SPRAY
SHT.04

FROM F.W. TE-075
SHT.06

FEEDWATER HEADER
SHT.06

DRUM LEVEL
SHT.01

DRUM LEVEL
SHT.01

TO ECONOMIZER
SHT.01

TYPICAL FOR BOILER NOS. 1B-100, 2B-100 & 3B-100

SOLID WASTE RESOURCE RECOVERY FACILITY

SINGLE/3 ELEMENTS BOILER
DRUM LEVEL CONTROL

Gibbs & Hill, Inc.
DESIGNERS, CONSTRUCTORS
New York

SCALE: NONE

SH. NO.
02

B1C-0203

JOB NO.

FILM NO.	ISSUE	DATE	DWN.	CHKD.	ARCH	STR.	MECH	ELEC.			AGD		SM		HCS	CONFIGURATION
	A	5/11	RLR	AJS							AGD		SM		HCS	CLIENTS REVIEW
	O	10/76	RLR	AJS							AGD		SM		HCS	CONFIGURATION
MICRO																

APPROVALS   ISSUED FOR

173

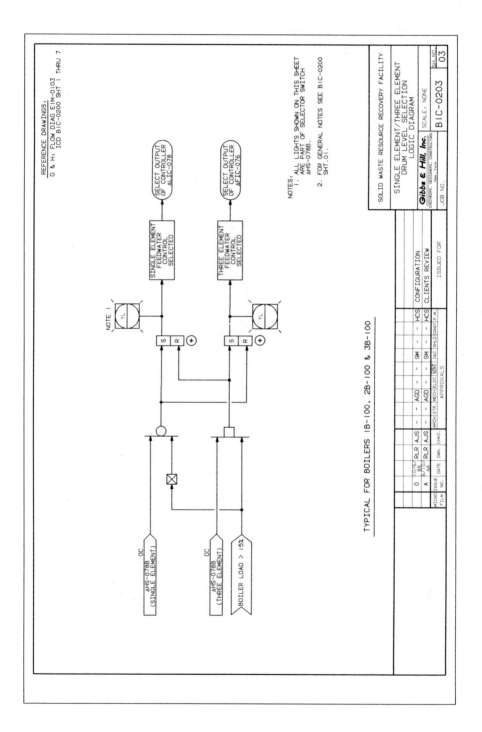

REFERENCE DRAWINGS:
G & H: FLOW DIAG E1M-0103
ICD BIC-0200 SHT 1 THRU 7

SELECT OUTPUT
OF CONTROLLER
ΔLIC-078

SINGLE ELEMENT
FEEDWATER
CONTROL
SELECTED

SELECT OUTPUT
OF CONTROLLER
ΔFIC-076

THREE ELEMENT
FEEDWATER
CONTROL
SELECTED

NOTE 1

YL

S R ⊕

YL

S R ⊕

NOTES: 1. ALL LIGHTS SHOWN ON THIS SHEET
ARE PART OF SELECTOR SWITCH
ΔHS-078B.

2. FOR GENERAL NOTES SEE BIC-0200
SHT.01.

OC
ΔHS-078B
(SINGLE ELEMENT)

OC
ΔHS-078B
(THREE ELEMENT)

BOILER LOAD > 15%

TYPICAL FOR BOILERS 1B-100, 2B-100 & 3B-100

SOLID WASTE RESOURCE RECOVERY FACILITY

SINGLE ELEMENT/THREE ELEMENT
DRUM LEVEL SELECTION
LOGIC DIAGRAM

Gibbs & Hill, Inc.
ENGINEERS, DESIGNERS, CONSTRUCTORS
New York

SCALE: NONE

BIC-0203

SH.NO. 03

JOB NO.

MICRO FILM NO.	ISSUE NO.	DATE	DWN.	CHKD.	ARCH. STR.	ARCH	MECH	ELEC	I&C	SHLD	CONST	HCS				CONFIGURATION
	O	10/6 AR	RLR	AJS	-	-	AGD	-	SM	-	P.M.	HCS				CONFIGURATION
	A	5/7 3R	RLR	AJS	-	-	AGD	-	SM	-	P.M.	HCS				CLIENTS REVIEW
							APPROVALS									ISSUED FOR

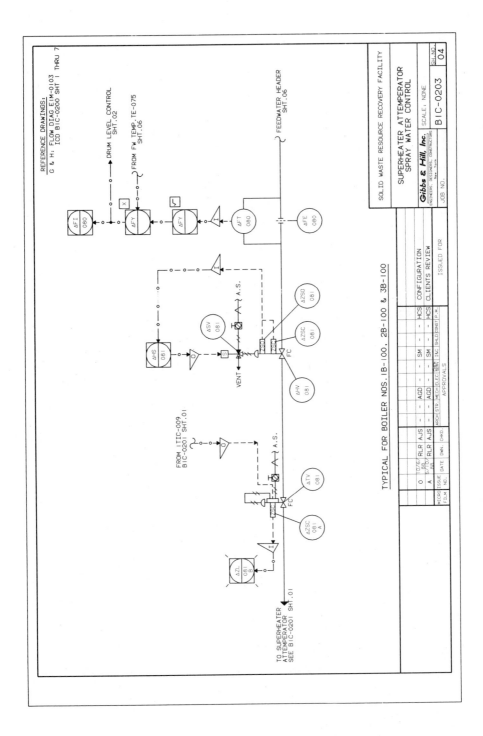

TYPICAL FOR BOILER NOS. IB-100, 2B-100 & 3B-100

REFERENCE DRAWINGS:
G & H: FLOW DIAG. EIM-0103
ICD BIC-0200 SHT 1 THRU 7

DRUM LEVEL CONTROL
SHT. 02

FROM FW TEMP. TE-075
SHT. 06

FEEDWATER HEADER
SHT. 06

FROM ITIC-009
BIC-0201 SHT. 01

TO SUPERHEATER
ATTEMPERATOR
SEE BIC-0201 SHT. 01

VENT

A.S.

A.S.

SOLID WASTE RESOURCE RECOVERY FACILITY

SUPERHEATER ATTEMPERATOR
SPRAY WATER CONTROL

Gibbs & Hill, Inc.
ENGINEERS, DESIGNERS, CONSTRUCTORS
New York

						HCS	CONFIGURATION
						HCS	CLIENTS REVIEW
O	10/6/98	RLR	AJS		-	AGD	-	SM	-
A	5/10/98	RLR	AJS		-	AGD	-	SM	-
MICRO	ISSUE	DATE	DWN.	CHKD.	ARCH	STR.	MECH	ELEC	HTG	INST	SVLD	CONST	P.M.
FILM NO.							APPROVALS						ISSUED FOR

| JOB NO. | BIC-0203 | SH. NO. 04 |
| | SCALE: NONE | |

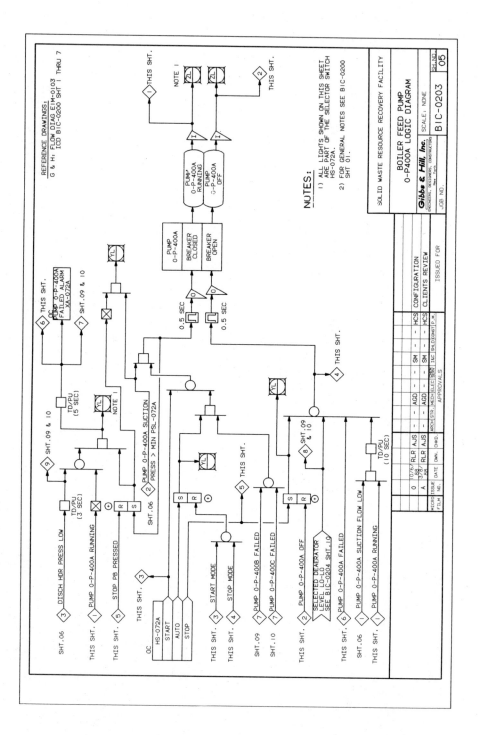

Index

Index

ABOUT THE AUTHOR

Michael J. G. Polonyi has over 17 years' hands-on experience in the design, operation, maintenance, and troubleshooting of power plants, electric energy systems, load dispatch, process control, power frequency control, reactive-voltage control, boilers, dynamic simulation, identification and optimization, data acquisitions systems analysis, installation, and start-up. He has a Master's degree in Electrical Engineering from Columbia University, has written extensively in trade publications and journals on the subject of power and process control, and has consulted for such organizations as U.S. Steel, General Electric, and New York University. Michael Polonyi is a resident of New York City.